Cybersecurity Simplified

A Beginner's Guide

by Gary S. Miliefsky

Legal Disclaimer and Copyright Notice

The information provided in this book, *Cybersecurity Simplified,* is for educational and informational purposes only. While every effort has been made to ensure the accuracy of the information contained herein, the author and publisher make no guarantees or warranties regarding the completeness, reliability, or applicability of the concepts, strategies, or recommendations presented.

The PANCCD™ Model and related methodologies are not substitutes for professional cybersecurity consultation, legal advice, or compliance guidance. Readers are encouraged to consult qualified cybersecurity professionals, legal advisors, or other experts to address specific risks, legal obligations, or technical requirements within their organization.

The author and publisher expressly disclaim liability for any damages, losses, or disruptions caused by the implementation or reliance upon the information contained in this book, including but not limited to financial losses, data breaches, or regulatory non-compliance.

Any mention of third-party products, services, or organizations does not constitute endorsement or recommendation. The reader assumes full responsibility for decisions made based on the material presented in this book.

The PANCCD™ Model is a conceptual framework provided as guidance. The reader's unique circumstances and organizational context should drive its application and customization.
By reading this book, you agree to hold the author and publisher harmless from any claims or liabilities arising from the use or misuse of the information provided herein.

For permission requests, write to the publisher at:
garym@cyberdefensemagazine.com

Trademarks
The PANCCD™ Model and related marks are trademarks of Gary S. Miliefsky. Unauthorized use is strictly prohibited.

First Edition
November 24, 2024

ISBN: 978-1-966415-99-2

Published by:
Cyber Defense Media Group
1717 Pennsylvania Avenue NW, Ste 1025
Washington, D.C. 20006

https://www.cyberdefensemediagroup.com

Printed in the United States of America

About The Author

Gary S. Miliefsky is a globally recognized cybersecurity expert, entrepreneur, and thought leader. With decades of experience in the cybersecurity industry, Gary has been at the forefront of innovation, helping organizations navigate the rapidly evolving threat landscape with actionable insights and practical solutions.

As the creator of the **PANCCD™ Model**, Gary introduces a revolutionary framework that simplifies cybersecurity into six essential pillars: People, Apps, Networking, Computing, Code, and Data. This model empowers organizations to build resilience, enhance compliance, and proactively mitigate risks in an increasingly digital world.

Gary is the founder and publisher of **Cyber Defense Magazine**, one of the industry's leading publications, online at https://www.cyberdefensemagazine.com, where he continues to share his expertise and advocate for cutting-edge cybersecurity practices. He is also a sought-after keynote speaker, regularly appearing at global events to educate and inspire audiences about the importance of staying ahead of cyber threats.

Throughout his career, Gary has advised government agencies, Fortune 500 companies, and startups on developing robust cybersecurity strategies. His passion for making complex topics accessible has earned him recognition as a trusted voice in the cybersecurity community.

Gary S. Miliefsky founded **CyberDefenseCon** located online at https://www.cyberdefenseconferences.com with a mission to honor and appreciate the tireless efforts of Chief Information Security Officers (CISOs), who serve as the underappreciated first responders in the battle against cyber threats. Recognizing the immense responsibility CISOs shoulder in protecting organizations from evolving threats, Gary envisioned CyberDefenseCon as a platform to celebrate their achievements, provide them with valuable insights, and foster collaboration among the cybersecurity community.

Through this annual event, Gary has created a space where CISOs are recognized not only for their technical expertise but also for their unwavering dedication to safeguarding critical systems and data. CyberDefenseCon reflects his commitment to elevating the role of CISOs, promoting innovation, and building a stronger, more resilient cybersecurity ecosystem.

When he's not working on cybersecurity initiatives, Gary enjoys spending time with family and friends, working with and helping US Veterans and is committed to advancing awareness and innovation in the field.

This book, *Cybersecurity Simplified* focusing on the PANCCD™ model, is a reflection of Gary's mission to make cybersecurity simple, practical, and effective for businesses and individuals alike.

Contents

Acknowledgements

This book, *Cybersecurity Simplified,* is the result of countless hours of effort, collaboration, and inspiration. I am deeply grateful to those who have contributed to its creation and supported me throughout this journey.

To my family and friends, thank you for your unwavering encouragement and understanding as I dedicated myself to this project. Your belief in me has been my greatest source of strength.

To my colleagues and mentors in the cybersecurity industry, your insights and expertise have profoundly shaped my understanding and approach to this dynamic field. Special thanks to those who shared their perspectives and experiences, which have enriched the content of this book.

To the readers of this book, whether you are seasoned professionals or just beginning your cybersecurity journey, thank you for trusting me to guide you. It is my hope that the PANCCD™ Model becomes a valuable tool in your efforts to protect what matters most.

Lastly, I want to acknowledge the countless cybersecurity professionals around the world who work tirelessly to defend our digital lives. Your dedication inspires me every day and reminds us all of the importance of vigilance, innovation, and collaboration.

Thank you all for being part of this journey. Together, we can make the digital world a safer place.

Foreword

In an era where digital transformation drives the way we live, work, and interact, cybersecurity has never been more critical—or more complex. Every day, organizations of all sizes face new challenges, from sophisticated ransomware attacks to the subtle infiltration of insider threats. Yet, amidst this evolving threat landscape, one truth remains constant: resilience is not an option—it's a necessity.

Cybersecurity Simplified arrives at a time when clarity and actionable frameworks are desperately needed. Designed to break down the intricacies of cybersecurity into six digestible pillars— People, Apps, Networking, Computing, Code, and Data—the PANCCD™ Model empowers individuals and organizations to navigate an overwhelming world of threats with confidence and purpose.

This book is not just another theoretical exploration of cybersecurity concepts. It's a practical guide written for everyone—from board members and executives seeking to understand their organization's vulnerabilities, to IT professionals striving to implement robust defenses, and even newcomers eager to grasp the fundamentals of digital safety. By focusing on simplicity and practicality, the PANCCD™ Model provides a pathway to resilience that anyone can follow.

What sets this book apart is its actionable nature. Each chapter is filled with real-world examples, straightforward strategies, and insightful guidance that can be applied immediately. It's not just about identifying risks—it's about understanding how to address them effectively and sustainably.

As you turn these pages, you'll find yourself equipped not only with knowledge but with the confidence to make informed decisions in the face of uncertainty. Whether you're securing your business, protecting personal data, or looking to enhance organizational culture, this book is your blueprint for success.

I encourage you to embrace the principles and strategies outlined here. In a world where the digital frontier is both an opportunity and a battleground, *Cybersecurity Simplified* is a timely and invaluable resource for staying one step ahead of the next breach.

Sincerely,

Gary S. Miliefsky, CISSP®, fmDHS
Publisher, Cyber Defense Magazine
https://garymiliefsky.com/

An Introduction to Cybersecurity

In today's overwhelmingly digital world, cybersecurity is a constant concern, organizational challenge, and indispensable part of life. It does not matter whether you are an individual who wants to protect personal data or a corporate representative eager to shore up weakness in existing security infrastructure. The same precise vulnerabilities and solutions exist at every level of today's digital existence.

As we grow more dependent on computers, smartphones, other connected devices, and the online world for activities of daily life and business, cybersecurity becomes ever more important. It encompasses a vast array of strategies and practices designed to safeguard you and your data. If you do not know what they are and how to implement them, you risk losing control of your most important and powerful asset: data.

This book will guide you through a unique model that explains security in a truly accessible and understandable way. By exploring each element from every angle, you will gain more in-depth knowledge about dangers and resolutions.

I have developed the PANCCD pyramid as a helpful guide to resiliency, protection, and risk management. Join me on this journey to empower yourself with the knowledge needed to protect your data and digital assets in this increasingly interconnected and vulnerable world.

What Is PANCCD?

The pyramid model displayed above simplifies the complex realm of cybersecurity for beginners. Experts who recognize these elements still need more information about the specific risks and possible solutions. Each chapter of this book will dive deep into further detail about the specific layers of PANCCD and guide you toward actionable solutions.

This acronym represents a model for understanding cybersecurity at every level:

P – People

A – Applications

N – Networking

C – Computing Devices

C – Code

D – Data

> **Tip: Start asking questions about cybersecurity risks and resilience using this simple model.**

Each of these areas represent potential security risks and opportunities to improve the strength and resiliency of your cybersecurity efforts. When an attack threatens your system or data or a successful breach occurs, it is usually associated with one of these elements.

PEOPLE — 1. Awareness & Training

APPLICATIONS — 2. Securing applications through development and testing.

NETWORKING — 3. Protecting data flow and network access points.

COMPUTING — 4. Securing devices and systems.

CODE — 5. Writing and maintaining secure code.

DATA — 6. Ensuring data integrity & availability.

GOALS

1. Building a security-first culture to prevent human error and insider threats.

2. Securing applications through development and testing.

3. Prevent unauthorized access and data breaches through network security

4. Maintain secure configurations and updates for endpoints and servers.

5. Ensure robust coding practices to minimize vulnerabilities

6. Protect data through encryption, access control, and lifecycle management.

As you read through this book 'Cybersecurity Simplified,' you will come to understand how the layers of the PANCCD pyramid function in the security realm, what types of vulnerabilities they introduce, and how to make changes to reduce or eliminate the risks.

People

The Human Factor: Strengthen Cybersecurity Through Awareness and Education

When many people think of cybersecurity, digital fortifications and cutting-edge technology come to mind first. Unless you are involved in the IT world yourself, these things may seem out of reach or totally foreign. Of course, these types of high-tech things are integral parts of the risk identification and solution process. It is important to remember, however, that people are one of the most critical parts of the whole complex equation.

The human factor affects security in many substantial ways. People's actions (or lack of action) can either improve or compromise the security of systems, devices, and data. All individuals – from IT team members to everyday product users – play a pivotal role in the ongoing battle against cyber threats.

People introduce many risks into any tech system they interact with. Things like awareness and education matter a lot when it comes to creating a cybersecurity model that's fast and easy to deploy and understand. However, you cannot eliminate human error entirely. When considering the dynamic intersection between people and technology, you can only do so much.

> ⚡ **Tip: Start by identifying and understanding the risks and threats. Then, create a strategic approach to minimize and mitigate them. Ultimately, you can create and implement best practices to reduce vulnerabilities and inspire more vigilant protection of the digital world by the people who rely on it every day.**

Identifying the Digital Risks and Threats

As people are omnipresent in any area having to do with computing, digital networks, and software use, they represent a wide range of risk factors.

The cyber realm involves a huge array of complexities, and the people who dive into this world are not always equipped to understand or deal with them properly. This leads to weaknesses that are difficult to overcome.

Digital risks and threats specifically related to the human element occur all the time. Some of these are perpetrated by criminals or hackers who want access to data. Unfortunately, some also fall under the responsibility of those who work for companies that need to protect it.

These are the most common human variable risk factors:

Falling for Phishing Attacks

As much as modern internet users get warned about never clicking on suspicious links in emails or text messages, they still do so with alarming regularity. One of the biggest issues is that people cannot identify what is suspicious and what is not. This is not a matter of ignorance or simply not paying attention. People who create phishing emails have many tricks up their sleeve to hide nefarious links or disguise URLs as reputable ones.

Phishing is a technique that digitally 'fishes' for information or access directly. Malicious people or systems pose as legitimate entities and manipulate people into giving up sensitive data or allowing them to access it in other ways. This can include usernames, passwords, payment information, or other details that no one should have but reputable companies.

As cyber threats get more sophisticated, the phishing messages sent look increasingly real. They come from seemingly genuine URLs, have company logos and letterhead, and often include worrisome threats that people feel compelled to act on.

Phishing is often considered a social engineering threat. It uses psychological manipulation in order to reach a desired outcome. In other words, the criminals prey on people's fear, excitement, sense of responsibility, and other emotional responses. A person who gets an email from their boss asking them to set up an appointment to discuss a raise will eagerly click on the link to do so without thinking it is fake. Someone worried about finances will log on with their bank username and password if a message that looks genuine tells them that their account is at risk.

Weak or Repeated Passwords

While some systems or online networks require complex passwords with both capital and lower-case letters, numbers, and symbols, many do not include this important security feature. People who struggle to remember complex codes often opt for simple and easily hackable ones instead.

Among the most common passwords used anywhere include such cybersecurity failures like:

- 123456 (or any combination of sequential numbers)
- Admin or user (usually because the user didn't bother to change it)
- Password or P@ssw0rd
- Unknown or *****
- Simple combinations like abc123 or 102030 or axbxcx

Only slightly better are those that include the company name, initials of the user, or easily accessible information like birthdates or locations.

A related risk involves people sharing passwords with coworkers, friends, and family.

Using the same password for multiple accounts is another real risk. It makes remembering them easier, but also introduces greater risk. If a cyber attack happens that affects one account, the criminals then have a way to access other ones.

Failure to Update or Patch Software

Any time individuals are responsible for updating their own systems and software, some will ignore the notifications or put it off until later. They can take a long time, after all, and get in the way of work or entertainment pursuits. This leaves things open to cybersecurity risks and data breaches. Updates exist for a reason, and many of those have to do with identifying and countering emerging threats.

Every application, operating system, and software component get updates until they are no longer supported by the developers or ownership company, and they are deemed obsolete. These updates address known issues, bugs, glitches, and viruses. Cyber criminals target these weaknesses. They are very good at discovering them and sneaking in before updates close the door. The longer you put off patching programs, the greater the risk.

Not Protecting Physical Access

In both personal and professional settings, if you allow other people to access your devices, networks, or digital files, you run the risk of having the data stolen or tampered with. This is a physical protection problem. A person who leaves their work computer logged in while they go to lunch invites trouble. If someone brings their personal phone on a jobsite and uses it to access client or vendor data, the proper security measures will not be in place.

In corporate settings with on-site IT centers or servers, security measures must include limiting access to the physical location. You cannot just let anyone wander into these sensitive areas. Doors should have automatic locking mechanisms and multi-factor or biometric authentication systems in place. That way, only those who are permitted to gain access can do so.

Improper Data Handling

Many people simply do not know how to handle the information they are responsible for properly. While the organization may have protocols in place and automatic encryption, for example, security still depends on individuals following the rules. These issues arise due to both ignorance and negligence. It takes effort to tick all the boxes for a specific data handling method. People frequently want to cut corners or save themselves time and effort.

One of the common issues under this heading is unwitting disclosure or sharing of sensitive data. For example, someone missed a specific detail for a sales order, so they ask their friend in another department to send the information. Rather than doing it through proper channels, they email it, which is a less secure way of doing things.

Human Error

Mistakes happen. The majority of security issues caused by people within an organization have nothing to do with outright negligence of malicious activities. People are not perfect, and they may unknowingly do something that weakens security for even a short period of time. That is enough for bad actors to take advantage of that vulnerability.

No matter what the risk factor involved from a human perspective, there are ways to counteract the issue before it becomes more serious. First, the organization must implement strategic mitigation policies. Training helps reduce errors but can never do away with them completely. A quality and quick incident response plan also helps. Only then can they make their cybersecurity efforts truly bulletproof.

A Strategic Approach to Risk Mitigation

Once you identify human factors that create cybersecurity problems, you must create an effective risk management strategy. Technological advancements play a crucial role in fortifying the defenses. However, true success comes down to the efforts of the people who create the systems and ensure that everyone else follows them.

People are, after all, not only potential vulnerabilities. They are also the key defenders against cyber threats. While human behavior can introduce a lot of risk factors, you also need to create a strategic plan to make them valuable allies. Technology takes up the slack and overcomes these weaknesses sometimes, but it is possible to bolster knowledge, adaptability, and decision-making skills to align them with robust security protocols at the same time.

Getting Your Priorities Straight

Before making any huge changes across the board, dive into past and current weaknesses and figure out where your organization needs help. It also helps to stay up to date about things happening in the industry. Look up what the most common issues are, such as those listed above, and figure out what others do to minimize them. What are the priorities when it comes to decreasing the human risk element?

Conduct full risk assessments that look into how people interact with the technology and networks, any discrepancies, dangerous behaviors, and even places where bad habits have been allowed to flourish. This can include things like security awareness surveys, phishing susceptibility tests, and quantifying human errors.

While you may need to completely overhaul your cybersecurity efforts in the end, it makes more sense to start with the most immediate issues. Once these priorities are handled successfully, you can aim for bulletproof perfection.

Technical System Security

One of the best ways to reduce issues caused by human error is to automate things like updates and patches and take as many things out of people's hands as possible. As the world becomes increasingly digital and uses more smart technology, why not take advantage of the options for your overall security strategy? Not only does this reduce mistakes, but it also simplifies everything, so it is easier to check on, update, and improve when problems arise.

Technology improvements stand at every level of the cybersecurity strategy. Today's smarter AI and ML systems present new opportunities to strengthen things even more. You can employ proactive threat detection and responses based on predicted vulnerabilities and past data about people-led issues.

Tech simply makes it easier for human failings to have a much smaller impact on the entire organization.

Grow a Top-Down Culture of Safe Habits

C-suite executives are no less likely to choose weak passwords or click on real-looking links in emails than administrative staff in the smallest department. In order to make changes that get ahead of the next breach, the strategy must include everyone. This includes managers and project team leaders, mail room assistants, established employees, and brand-new hires.

Company culture makes a huge difference in the war against cyber criminal activity. If certain departments or levels of employees get left out of the system, they are much less likely to comply with implemented protocols or report errors or issues as they arise. Keep lines of communication open and make sure everyone knows they can and should notify managers or the incident response team representatives without fear of repercussions.

Now that you understand the need for a comprehensive strategy to deal with human-related security issues, the next step is to develop an effective plan. This may look different for every organization. However, the following section outlines some vital tips to help you improve your chances of avoiding cyber-attacks and data breaches.

Solutions to Reduce Human Security Issues

The time has come to take action. Identifying human risk factors at the top of the cybersecurity threat pyramid is only the first step. Creating a strategy to deal with them must include actionable steps and functional plans to eradicate issues before they get worse.

As mentioned above, people are an organization's greatest asset and a true help when it comes to shoring up security issues. The

'P' at the top of the PANCCD pyramid cannot be ignored if you want everything to get better and safer. Keep in mind that the power to stop future breaches is primarily in the organization's hands. You cannot rely on individual efforts to make improvements sufficient to counter the well-organized criminal behavior. The following changes to practices and protocols can make it happen.

Take Away the Human Risk

One way to remove people from the security equation is through automation and smart tech. As long as you can keep these systems locked down, they can certainly help reduce vulnerabilities across the board.

Some options include:

- Time-triggered computer shutdowns or lock outs that require a password to regain access. This means a person can never forget to secure their device when they head to lunch, a meeting, or home at the end of the day.

- Strong password requirements. Set up the network and apps to require more complex and longer passwords. Do not let people get away with hackable '12345' or 'password' as their log-on information anymore.

- Use multi-factor authentication. People who are allowed access must go through more than one layer of security. This commonly includes regular log-in information plus a code sent to their email or mobile device. Biometric authentication with fingerprint or retinal scanners adds even more protection.

- Email filtering and anti-phishing assistance. Set up in-house email systems with the latest automatic detection systems

designed to identify fake links and questionable URLs or
email addresses.

- Opt for update management systems. When all patches and
updates are installed automatically, they will always be in
place in time to stop many cyber-attacks. Make sure that
users cannot bypass or delay updates manually.

These tech-based solutions work together with other smart
security strategies and policies. They act as another layer of
protection between threats and the people who need it most.

Better Security Education and Training

When it comes to the people portion of cybersecurity, one truth
remains constant: knowledge is power. Understanding risky
behavior, potential issues, and what to do to mitigate them is the
most effective defense.

Part of an organization's strategy must include regular and
updated training throughout the employee life cycle. Never expect
that giving someone a few tips or rules once will keep their
defenses up all the time. Build vigilance into the system.

Two options exist when it comes to security training. You can
develop an in-house program or outsource it to expert security
providers or consultants. Sometimes, developers of security
systems or programs have their own training provided.

The size of the organization and budgetary limits often define
which you choose. No matter what, you need to make sure it
covers all risks, issues, and best practices if you want it to make a
difference.

Before diving into the specifics of what makes security education
and training effective, understand that you should require
everyone to stay up to date and verified. This includes the C-suite,

assistants, tech workers, the sales department, field technicians, and folks that work in the lunchroom. They made need diverse programs crafted specifically for their role. Of course, the particulars depend on your company's make-up and employee positions.

What topics should you include in the cybersecurity training?

- Start with the basics. Many employees, especially those in older generations, may not understand what cybersecurity is and why it matters. Explain the potential impact of a data breach, and how it can affect not only their job, but their personal information as well. Lack of understanding leads to a lack of care.

- Insist on password security. Share details about how quickly a hacker can automatically break through the easiest passwords (less than one second for the worst!). Provide guidance on creating stronger, less vulnerable options that people can still remember.

- Teach phishing awareness and general internet safety. Do not click on random links. Do not share personal or business data through email or unknown online forms. When in doubt, go to the source for direct information.

- Establish the need for online privacy especially related to business details and data. Make rules about never accessing personal sites or social media through work machines. Also, teach people not to access work-related platforms from their own devices.

- Train everyone in incident response. What should they do if something goes wrong, or if they suspect a breach occurred? Make reporting things quickly a simple, straightforward process.

Overcoming Resistance to Cybersecurity Changes

Unfortunately, some people will be resistant to the idea of additional training or more complex protocols. This is especially true for people who are well-established in their jobs, come from older generations who are not as comfortable with everyday technology use, and those who do not value their position highly. Job satisfaction levels even come into play.

Lack of Awareness and Understanding – Insufficient knowledge about the importance of cybersecurity and the potential consequences of a breach can definitely contribute to training refusal. Ironically, it is only through that training that they can gain more awareness.

Inconvenience -- Some workers may see the new protocols as a waste of time or too burdensome. They do not want their established workflows disrupted.

Overconfidence – People who are sure that their personal abilities are good enough already may resist the entire concept of learning new things.

Fear of Change – Thos is a particular problem to people who entered the workforce long ago or who are already knowledgeable about past protocols. They do not feel comfortable with more complex security measures.

Skepticism – Employees may not trust their employers or the executives who are in control of things. They may also not believe that cybersecurity threats are so common or dangerous and consider them fearmongering or part of some wild corporate agenda.

Technology Intimidation – The average person not involved with the IT industry does not know much about the inner workings of any type of technology. If cybersecurity protocols seem too complex, they may find them intimidating and dismiss the idea that they could learn them.

Compliance Fatigue – When regulations, rules, policies, and strategies pile up on a single person, they may feel like they deal with more 'red tape' than actual work. This leads to them feeling overwhelmed and resisting additional requirements.

Addressing these issues is an integral part of getting all the people in the organization and outside stakeholders involved with the cybersecurity system. Aim to make things as user-friendly, accessible, and mutually beneficial as possible. These things foster a company culture of security focus and support.

People, the top of the PANCCD cybersecurity pyramid, represent many risk factors and stand as a prominent solution to an organization's vulnerabilities.

By creating a strategic plan, addressing problems, and taking a proactive approach, you can strengthen the defenses. Ongoing education and verification of best practices can help even more.

In the next section, we move down the pyramid to **Applications**. Dive into the inherent risk factors and common exploitations of software you may use every day.

Applications

Defending the Digital Gateways: Safeguard Software in a Threat-Filled Landscape

Applications or software programs of all types run the world these days. A vast array of options integrates with both personal and professional life. From mobile apps and web platforms to complex software running critical business systems, applications are the digital gateways through which we communicate, transact, engage in commerce, and access and process huge amounts of data every day.

With such a great amount of importance, it stands to reason that applications are one of the critical layers to the cybersecurity PANCCD pyramid.

To effectively safeguard all types of apps from a multitude of potential threats and vulnerabilities, you must first understand what they are. The sections of this chapter below will walk you through basic knowledge about the risks and how nefarious individuals, hackers, and automatic security breaching systems attempted to break through and get exactly what they want.

Application Vulnerabilities, Threats, and Hidden Risks

If you are not a software engineer or app developer, you may not understand the complexity of creating something that functions well with sufficient security to hit the market successfully. However, you should be aware that it takes a lot of work to create these things that people use every day. Some developers are more scrupulous than others. Cutting corners may occur particularly on simpler apps designed for entertainment or basic communication.

> **Tip: No matter what the application is, everyone involved with its creation and use should understand the vulnerabilities, threats, and risks of using it. While this list is not exhaustive, it does cover the most common issues that apps must protect themselves from to remain a strong part of an overall Cybersecurity system.**

Common App Vulnerabilities

These things are problems that are worked into the software itself. This is primarily an issue for developers and people who maintain the apps and create patches or updates. Some of these are called zero-day vulnerabilities, which are ones unknown to the people who developed the apps in the first place. They exist and can be taken advantage of on the first day (or Day Zero) of usage. Some software is simply unsafe from the start, and teams who release these products may not have the time, ability, budget, or interest in also releasing security updates or patches.

Security gaps and vulnerabilities come in a variety of styles. These are a few of the risks that are far too commonly built into the software applications themselves.

Cross-Site Scripting (XSS) – This type of vulnerability affects webpages. They allow attackers to insert dangerous or data-stealing scripts, which are then used or activated by unknowing people who visit the site.

Cross-Site Request Forgery (CSRF) -- If an attacker can manipulate the app or webpage in order to trick the user into doing something on a site where they did not intend to, it can be a serious cybersecurity risk.

Cryptographic Failures – Apps of all kinds transfer a lot of data back and forth with the user. If this is not encrypted properly, it becomes exposed to people who can use it for their own personal gain. This is one of the most common issues in data breaches related to identifying information, health records, credit card numbers, and similar things.

Security Misconfiguration – When the teams who create the app in the first place fail to properly configure the system, cloud service permissions, security features, or any other human error issue (such as default administrative usernames and passwords), attackers take advantage and sneak in.

Insecure Authentication – One of the simplest and most obvious potential app vulnerabilities has to do with inadequate methods for logging in or otherwise gaining access to information or the software itself. This is more easily exploited by people with nefarious intentions. This vulnerability is related to broken access control and covers everything having to do with unauthorized use and administrative powers.

Third-Party Dependencies – Any piece of software that relies on a third-party application or system can suffer from vulnerabilities related to that connection or the other app itself. Even developers who regularly update and maintain their own creation have no control over what the other team is doing or failing to do. This is a big problem when the third-party app uses obsolete technologies or frameworks.

Application Security Threats

While any vulnerability is obviously an issue for the people that use apps or want to keep systems and information secure, it helps to understand the specific threats in order to prevent them. These are the methods that attackers use to create vulnerabilities or exploit them.

Exploiting Weaknesses – The Art of App Attacks

In the realm of criminal and malicious activity, there are always new methods of trying to access apps, exploit vulnerabilities, and get at the data or systems that the attackers can use for their own purposes.

Forget the Hollywood image of a single hacker typing away in their basement trying to sneak into a mainframe or siphon money out of a bank account. Today's cybersecurity threats are vast, organized, and highly automated. They know how to find zero-day vulnerabilities, emerging gaps in protection, and take advantage of them all very quickly.

When you understand the inner workings of these types of attacks, it becomes easier to create a protection strategy. Of course, the average person can do little about these things other than to avoid out of date apps and remember to scan for malware and similar issues.

Companies must do more to protect their assets and employees.

Developers and security professionals who work independently or for a specific company need this information in order to build robust defenses against the ever-evolving threats. Unravel the mechanics behind various application attacks and illuminate the art behind the efforts.

From manipulating input data to sneaking into databases, hackers know the best methods. It is something that you need to understand as well, if you want to protect digital gateways, fortify data protection, and protect yourself and your business from exploitation.

How Cyber Attackers Exploit Vulnerabilities

This is a small collection of the methods used to bypass app security features and overcome weaknesses. These are the things that developers and users must find ways to protect against in order to safeguard systems and information.

Injection Attacks – Malicious code is inserted into input fields in the app in order to access the database or system and manipulated for nefarious purposes. This could be an SQL injection or something similar. Three specific vulnerabilities that fall under this heading include:

- *Local File Injection (LFI)* – This technique tricks apps into revealing hidden files by convincing the system that the added code is a trusted part of the original. It is the umbrella term for the following specifics.

- *SQL Injection* -- This method for exploiting weaknesses in an app's SQL database uses natural user behavior, such as executing a search query, to sneak in through vulnerable spots and gain access to data that should remain hidden. It allows the attackers to create or change user permissions, steal or manipulate data, or destroy it altogether.

Distributed Denial of Service (DDoS) Attacks -- Multiple computers or botnets flood a target system, network, or app platform with a massive volume of traffic. This overwhelms their resources and frequently causes things to shut down for a period of time. While disruption of services does not usually compromise data, it can leave users frustrated and the owners scrambling to fix things.

Malware Infections – There are many ways that malware gets into a piece of software. Many of them have to do with user behavior. People download unprotected or unknown files, and their devices get corrupted, which allows the malware to spread. Attackers can also upload these things directly if the app does not use sufficient input validation.

The incredibly long list of cybersecurity vulnerabilities, weaknesses, and methods to exploit them all need a robust strategy to combat any nefarious action. These types of things are often included in the development stage, especially when it comes to ongoing development for updates and patches.

Also, a large part of the responsibility falls on the user. They need to avoid the types of things listed in the "People" chapter of this book. These include weak passwords, inappropriate use of a program or platform, and falling for phishing scams.

From a business or organizational perspective, the only way to create bulletproof app security is to formulate a functioning strategy for the safeguarding process. Besides appropriate training for users, the following strategies will help you protect apps throughout their entire lifespan.

Strategies for Safeguarding Software

In this digital age, apps of all kinds have become the backbone of both personal and professional operations. Taking steps to create a functional strategy for safeguarding them against cyber threats is of the utmost importance. Develop proactive approaches. Learn best practices. Put robust methodologies into place for developers and the organization at large. All these things must work together to ensure the integrity and security of every software application.

This book does not teach software product development or coding. Instead, it delves into a comprehensive array of strategies that can fortify digital gateways and protect your interests. Every suggestion will not work for every app or company. Your team should know and understand the specific vulnerabilities and weaknesses that could cause problems down the road. This way, they can better prepare for any type of attack.

These days, the flood of cyber threats is virtually endless. The most important thing to do is develop protocols and framework for development and ongoing app support that builds resiliency and reliability into every digital asset. Here are some methods that may work well in your overall software safeguarding strategy.

Secure Coding Practices

High-quality coding and programming practices lie at the heart of all app protection practices. This has to be the foundation of your overall strategy for creating apps and making sure you choose the right ones to use as well. Make sure your company's developers know what they are doing. If you are not involved in digital product creation, only choose applications from reputable brands.

More information about coding and prioritizing security throughout the development process will come later in this book in the Code chapter. This is the second C in the acronym PANCCD.

Web Application Firewalls (WAFs)

Most computer users know that a firewall is very similar to the architectural feature of the same name. It forms a boundary between things that block unwanted problems from accessing them. In an apartment building, a firewall helps make sure that a conflagration does not spread from one housing unit to the next. In the world of software, a web application firewall also acts like a block or fortification designed to stop and neutralize malicious forces.

Your overall app security strategy should include frequent and effective use of firewalls that are designed to scan and filter incoming traffic and data input. This makes it possible to catch nefarious requests and cyber-attacks before they can get all the way inside. They do this by analyzing HTTP requests and responses through the use of predefined rules and policies. These are associated with some of the common vulnerabilities listed above.

Web application firewalls are a necessary part of an overall strategy to protect every app. They represent a proactive approach so that attacks cannot gain a foothold in the software or system.

Regular Security Patching and Updates

It is impossible to create an app that will always remain as secure as it was at the start. Attackers are constantly looking for new ways of infiltrating software and systems in order to take control and steal data. If you do not keep the software updated and patched to counteract these threats, you will fail in your efforts to create Cybersecurity Simplified.

Security patches and updates represent a critical line of defense against all types of vulnerabilities. They are one of the best ways of taking a proactive stance against future issues. Regular updates also make the apps more valuable, functional, and appreciated by end users. It is a win-win practice.

There are a variety of different types of updates that should exist for any type of app. They can include operating system patches, firmware updates, and updates for the software code itself. Developers should stay on top of emerging issues and frequently test the apps for anything they missed the first time around. Users must apply the updates immediately when they become available, or you risk suffering from one of those human-caused vulnerabilities listed earlier in the book.

Implementation of Strong Access Permissions

If you can keep malicious people or bots out of your app, it greatly reduces the risk of anything unwanted occurring. Of course, this is not 100% possible no matter how strict your cybersecurity strategies and practices are. One way to counteract vulnerabilities is to create a very strong system of access permission. In other words, make sure only people you want to use it can get in and do so.

The simplest methods to do this involve things like the forced creation of complex passwords and unique usernames. It can also include multifactor authentication or even biometric authentication like fingerprint or retinal scanners. Likewise, control who has access to certain systems or data carefully. Make sure all employees understand that they should never grant permissions to coworkers or anyone else even if they outrank them in the business hierarchy.

Security Monitoring and Incident Response Rules

Another important part of your overall cybersecurity strategy includes logging activity related to general use and data access. The more information the security team has about who accessed what and when, the easier it will be for them to take effective countermeasures. Build some type of clear alert or notification protocol into this system. It should be easy for the app user to recognize if something goes wrong so they do not potentially cause a greater security issue with their actions.

Alongside this type of monitoring and notification, it is also important to develop incident response rules. These must clearly tell the user what they should do as soon as they either do something wrong themselves or notice some unusual or unwanted response from the app. This is the type of thing you should train members of your team for, and it helps to give a written explanation or some sort of digital alert that tells them what to do.

Ongoing Assessment of Third-Party Assets

If your app relies on any third-party platforms or systems, make sure you check them frequently to qualify their security status. Attackers can sneak in a backdoor through an insecure app that is connected to yours. Unfortunately, this may require a complete overhaul of how the software functions or what it relies on for certain aspects of the features that users enjoy.

Organizations frequently rely on outside assets like APIs, frameworks, or libraries. These can greatly increase functionality and efficiency. However, they also run the risk of introducing weaknesses. Not only must you conduct due diligence when choosing the assets in the first place, but you must also monitor them for updates or alerts that are no longer supported by their developers.

Frequent Testing and Scanning

As mentioned before, things change quickly in the cyber threat realm. You must move even faster than the attackers if you want to protect the apps from issues. Continuous assessment is an essential part of any security strategy. There is no such thing as "set it and forget it" in the realm of app development.

A variety of different testing methods can help. These include vulnerability scanning, penetration testing, threat modeling, and more. It is important to know how the app will respond to a specific type of threat or attack so that you know how to patch it to minimize the risk.

All these strategic methods and more should already exist within your organization when you launch an app or adopt its use for in-house purposes. They help maintain the highest level of security at present. However, it is essential to include the idea of future proofing into your strategy. New tech emerges all the time. As you adopt more robust programs, so do the attackers who want to gain access.

Future Proofing – Proactive Tips for Emerging Threats

New threats and attacker techniques appear all the time. These frequently match the emergence of new technologies that catch the attention of app developers and software product design teams. Today, smart devices, artificial intelligence (AI) and machine learning (ML) have rapidly grown in popularity. This makes them prime focus points for people and groups who want access to your apps and data.

So many elements go into cybersecurity for the continuously evolving world of applications. Whatever strategies you devise must have room for these emerging issues. Future proofing projects leads to a brighter, more secure future. Interestingly enough, many cybersecurity experts are looking to new technologies to augment existing and planned measures.

Embracing AI and ML in Cybersecurity

It is important to understand AI and ML app risks systems become more popular. However, people in the cybersecurity world must also explore the opportunities that they can use for threat detection and rapid response. By implementing these algorithms into things like employee behavior analysis, anomaly detection, and reporting compliance, it is possible to have a faster and more efficient overall security response.

Securing IOT Apps and Ecosystems

The Internet of Things and smart gadgets have become much more popular in recent years. Every self-controlled thermostat, lighting system, or other device involves a control app that people use to set things up and keep things running smoothly. No one wants their home security or environmental comfort systems to fall under the control of a nefarious attacker. There is also the risk of security breaches through these types of apps that can compromise personal information.

Potential Threats in Virtual and Augmented Reality

As the line between the digital world and real life grows ever thinner, cybersecurity professionals need to investigate options for making AR and VR experiences safer. While there are many complex technological truths in these sectors, the apps that allow users to enter new worlds need the same type of features to mitigate risk.

There is a high degree of human error or negligence issues related to them, of course. However, for app developers and businesses associated with these types of future technology, it helps to integrate security considerations into the development lifecycle and keep abreast of the need for frequent updates.

This second layer of the PANCCD cybersecurity pyramid, apps, represents an important part of the entire equation. There are many intricacies involved with the practice of safeguarding software against a wide variety of threats and vulnerabilities. You need to understand the fundamental principles of secure development and embrace the ongoing needs of frequent testing, updates, and patches.

Web application firewalls, more robust access controls, and testing of third-party assets are all a part of app security. No matter what type of software you or your company is involved with, it is important to understand that the tech landscape constantly evolves. You must embrace a proactive mindset and prioritize security at every stage.

The next layer down in the PANCCD cybersecurity pyramid involves **networking**. While apps may serve as digital gateways for interactions, communication, transactions, and taking advantage of weaknesses, networks formed the infrastructure that connects these gateways.

Networking

Cyber Fortifications: Strengthen Digital Highways Against Attackers

The third layer of the PANCCD security pyramid focuses on networking and network security for both in-house and cloud options. Securing software and training people to use it responsibly is not enough to create the highest level of protection possible. Everything is interconnected these days, and the networks that form these connections need attention too.

Networks allow data to flow seamlessly between systems, platforms, and programs. They need to allow this transfer of information and assets while at the same time blocking any unwanted intrusion or access. Due to their omnipresence in both business and personal computing, they play a vital role in safeguarding digital infrastructure and the people that use it every day.

> **Tip: To perfect cyber resiliency for networks, people responsible for these things must understand the foundational principles and best practices.**

First, it makes sense to recognize what the weaknesses, vulnerabilities, and potential exploits are. Only then can you mitigate risk, fortify connections, and prevent attackers from gaining access. This is not a one-time action when the network is set up in the first place. Instead, like all other cybersecurity practices, it takes ongoing effort, frequent checks and testing, and a proactive approach to maximize results.

Potential Exploits Against Networks

Whether you use technology heavily in your business or career or simply do so for personal purposes, you access an amazing array of networks all the time. Each one needs the right type of security in order to keep you and your data safe. IT professionals responsible for this should be aware of the myriad vulnerabilities and threats that can compromise the safety of these pathways.

Cyber criminals look for weaknesses that they can exploit to breach network defenses, disrupt operations, and get their hands on sensitive information. Knowing the threat is the first step in combating it.

Physical Network Vulnerabilities

Although much less common than virtual weaknesses and attacks, it is still important to physically protect the hardware and tangible infrastructure supporting the network. This includes servers, data centers, wiring closets, and any endpoints or access control items. In any professional setting that uses a physical network contained on-site, the people planning the security system must think about actually limiting access to people who should not have it.

This can include something as simple as locking the door to the server room. However, it makes sense to use more technologically advanced systems like keypad or thumb print entry panels. Always have surveillance equipment like cameras installed to keep an eye on who is coming and going. Also, protect endpoint hardware like routers, switches, and even the desktop devices used by employees. Put everything possible in place to prevent physical tampering or any type of nefarious monitoring if you want to protect the confidentiality of the network itself.

Hardware issues are another example of physical weaknesses. Out-of-date routers or other equipment can cause problems. It is important for the IT department to update firmware to patch vulnerabilities and replace hardware when it no longer serves its purpose well.

OS or App-Based Network Exploits

The vast majority of attack methods include access from a distance through technological means. These are perpetrated by malicious individuals or groups who want to create an entrance into a network that gives them immediate or ongoing control of the network or access to data. These are commonly called network intrusions.

The reasons why these network exploits work include the same issues that relate to app-based security problems:

- Outdated software that has not been updated or patched

- Lack of sufficient or correct firewall configuration

- Ignoring default security features in operating systems

- Unauthorized device access such as personal gadgets, printers, etc.

- Insufficient access rules or poor authentication protocols

Some of the potential threats include:

Man-in-the-Middle (MitM) Attacks – This general term covers cyber-attacks in which the nefarious actor infiltrates communication or data transfer between two or more authorized users or a single user and the app itself. Picture it like someone eavesdropping on a conversation by standing nearby. These often result in theft of log-on information and passwords, changed access permissions, and data breaches.

Network Port Scanning – Every point of access for a network is called a port. Cyber criminals scan for open or poorly protected ones all the time. They can then take advantage of the weak points in security to gain access or implant malware. These are things usually protected by firewalls.

Packet Sniffing – All data sent over a network is formulated into packets, which are sometimes called blocks or segments. These are the smallest unit of collected data transferred in this way. As the name suggests, attackers use sniffer tools to find and gain access to the data. These are also used regularly for beneficial purposes to examine networks and make sure everything is running smoothly.

DNS Spoofing – DNS stands for Domain Name Server and operates to make sure that approved users connect to the right networks, IP addresses, or URLs when they enter the appropriate information. Spoofing DNS involves the manipulation of these records in order to redirect people to malicious locations.

VLAN Hopping – VLAN or virtual local area network hopping involves sending data packets to ports not usually used or accessible. This is a way to attack the host in order to sneak into network locations that no one or a limited number of users should have access to. Once cyber criminals are in, they can move more easily throughout the network to reach their end goal.

Malware – This covers a wide range of malicious software types including viruses, trojans, keyloggers, worms, bots, ransomware, and rootkits. They are either directly injected into networks through weaknesses or come in through phishing and other social engineering attacks.

Some of these are much more common than others. They also represent only a selection of the potential network vulnerabilities that bad actors make use of. In the ever-changing world of digitalization, connectivity, and related cybersecurity, there are always new threats, methods, and hopefully solutions.

Social Engineering Attacks

This brings things back to the human element, which is always a weakness to consider when it comes to cybersecurity. Social engineering is about using human psychology and its related weaknesses in order to gain access to a network. This tactic manipulates people into giving away sensitive information, taking actions that lead to problems, or otherwise compromising security without any knowledge of the risk at all.

Cyber criminals frequently use things like phishing emails, impersonation of people within an organization or a related governing body, or even scam phone calls for these purposes. No matter what, social engineering is designed to pray on trust of authority, fear of personal repercussions, curiosity, and the common human hunger for any type of reward.

The best defense against these types of network vulnerabilities is training and education. There is simply no way to work against the natural psychology of people without telling them that they are themselves a risk factor and must remain vigilant against these types of scams.

Fortifications to Strengthen Tech Pathways

IT professionals and organizations need to employ intelligent protocols in order to fortify their network infrastructure against all types of vulnerabilities and risk factors. Cyber threats evolve and proliferate rapidly, and they will continue to do so as tech options expand. Protecting these data pathways safeguards proprietary information, preserves business functions, protects reputations, and helps you avoid considerable financial losses.

There are many things that security teams can do to bolster the resilience of their networks and work against risks from every direction. The following options represent some of the more common and effective.

Network Access Control (NAC) -- Implementing network access controls help secure networks by enforcing strict access and authentication policies. This helps businesses make sure that only specific users at administrative levels are able to use the network and devices connected to it. NAC options also verify the identity of the devices and make sure they are allowed. It also creates more transparent visibility over who is connected, how, and what they are doing.

Multi-Factor Authentication (MFA) – Anyone who has ever logged on to any online platform or professional network has experienced multi-factor authentication in practice. It requires more than one form of verification to prove you are who you are and that you have permission to access the network. This can include such things as verification links in emails, one-time use codes sent to smart phones, and other methods. MFA helps block people without insider access from using someone else's logon information nefariously.

Security Information and Event Management (SIEM) – As pivotal components of modern network security infrastructure, these solutions improve threat detection across network devices, servers, and nodes. In general, they make the entire threat identification and response process much more efficient and effective. Many of them include industry-specific capabilities like log management and compliance reporting for different regulatory bodies.

Network Segmentation – If an attacker gains access to a network that operates as a singular thing, they will find it much easier to access any part of it that they want. This is where the power of segmentation comes into play. By dividing it into small, isolated segments, it becomes much easier to keep unauthorized individuals out. It minimizes the scope of data breaches or any other type of infiltration. This is frequently achieved through physical divisions using switches, virtual local area networks, and routers as well as the effective implementation of strict firewalls and a variety of access rules.

Intrusion Detection and Prevention Systems (IDPS) – It is virtually impossible to do anything about a hacker or a bot infiltrating the network if you do not notice their activities in the first place. Intrusion detection and prevention systems monitor things continuously to make sure that no malicious activities or security breaches occur. They analyze traffic and logs and immediately identify and notify any potential security threats. This is one of the most effective real-time options that make a difference for cybersecurity. These solutions can do things like block unrecognized IP addresses and quarantine suspicious data packets or devices.

Endpoint Detection and Response (EDR) – This is another detection solution that works with endpoint devices rather than software or human-factor elements. It focuses on desktops, laptops, mobile devices, and the servers themselves that could give clues about incoming threats. Most importantly, EDR solutions are capable of isolating compromised endpoints and can quarantine files or programs introduced to them before they gain access to the network at large.

Regular Security Audits and Testing – Nothing makes a difference for cybersecurity more than keeping up with all best practices and solutions. You can never stop at "good enough" or skip active monitoring simply because nothing bad happened yet. Auditing the networks and the security strategies is a necessary part of the entire process. Test things frequently to ensure their continued strength and efficacy against emerging threats.

Organizations have a diverse range of options when it comes to network security. These primarily focus on physical or on-site servers. However, more companies than ever before looked to the cloud for their computing power. The methods of establishing security remain primarily the same.

Data encryption plays a larger role, but things like access management, security monitoring, compliance with regulatory requirements, and continuous detection and protection protocols must be followed. Above all else, make sure you choose the right vendor who is reliable and trustworthy.

Specific Security Considerations for Cloud Computing

An increasing number of companies turn to the cloud for a wide variety of computing and network options these days. Instead of having on-site servers and built in software or infrastructure, the cloud provides many of those things. They include Infrastructure as a Service (IaaS), Platform as a Service (PaaS), and Software as a Service (SaaS). If your organization uses any of these things, it may take additional efforts to secure the networks.

Cloud network architecture exists mostly beyond your control. The whole collection of virtual servers, storage systems, networking infrastructure, and management interfaces all need their own security protections in place to prevent unauthorized access and data breaches. These primarily fall on the shoulders of the cloud service providers. That means that your main responsibility is researching them carefully and choosing ones with the best reputation.

Most of the same security challenges exist for cloud networks as for physical ones. These include insider threats, data loss and theft, malware intrusion, and compliance issues. Each cloud provider needs to clearly state what their efforts are and if they use a shared responsibility model that clearly defines the respective roles.

You cannot leave everything to them. Instead, perform strict due diligence before getting involved and then implement robust access controls, encryption mechanisms, network segmentation, and intrusion detection and prevention systems just as you would for a non-cloud system.

Security Needs a Continuous and Proactive Approach

As with any type of tech security measures, it is of the utmost importance that those responsible for network protection take a proactive approach to its safety. As mentioned many times in this book, things change all the time, and the world of technology and digitalization progresses rapidly.

It is not enough to set up one detection system or upgrade the firmware once. A simple lock on an in-house server door will not keep everyone out forever. Do not trust one cloud server option without double checking its security features down the road.

Every layer in the PANCCD cyber protection pyramid needs robust understanding, intelligent and decisive action, and continuous attention to maximize security across the entire tech world.

Whether you look at apps or networks, people remain a factor in the ultimate protection of data and systems. Now it is time to look at the hardware or **computing devices** that allow people to engage in business, communicate, and enjoy entertainment every day.

Computing Devices

Build Digital Armor: Systems and Strategies to Protect Gadgets and Gear

How do people access the networks that require so much security? What hardware and onboard software allows you to do all manner of business and personal things? Everyone in the modern world uses different types of computing devices every day. These indispensable gadgets and gear power your digital life, keep you connected to coworkers, business contacts, family, and friends. They come in a wide variety of sizes and styles these days, but one thing remains a constant.

> **Tip: You must protect your computing devices with the strongest digital armor possible to minimize cybersecurity risks and keep everything operating smoothly.**

Increased connectivity comes with a much higher level of risk. Threats of cyber-attacks and data breaches appear in every part of your life. They compromise your security and put your private information at risk.

Learn about the vulnerabilities, protection methods, and strategies that can help you secure your gear and gadgets from an IT security perspective.

Understanding Device Vulnerabilities

There are two primary sides to computing device security: hardware and software. This is very similar to the concepts of legal protection versus digital protection mentioned in the Network chapter. Since the actual devices that people use for work or pleasure or dependent on the operating systems and apps used to perform different functions, you cannot really separate one from another when speaking about cybersecurity.

Software exploits and vulnerabilities form the primary pathway for attackers and malicious bots to gain access to the devices themselves. However, hardware flaws also represent another critical weakness that can compromise their security. These include:

Design Flaws

These are things that the product designer or developers did wrong. Unfortunately, not every manufacturer focuses on bringing the most secure product to market. Flaws happen due to oversight, rushed timelines, budget constraints, or a lack of rigorous testing. Oddly enough, administrative access put in place for debugging and maintenance before the product hits the market can actually create a weakness down the road. Cyber criminals can take advantage of the hardcoded credentials and use them as a pathway around normal access controls.

Design flaws, and a wide variety of varieties and specific issues. IOT devices, for example, may have built in weaknesses related to Wi-Fi encryption protocols or other details regarding their connectivity. Any computing device may suffer from hardware component flaws that affect memory modules, processors, and more.

The problem with these types of issues in the cybersecurity world is that the end user or professional team responsible for updating or maintaining the computers cannot do anything about these issues. The best bet is to work with reputable manufacturers and make sure the devices are well-designed a from the start.

Manufacturing Defects

Another element of the cybersecurity equation that is mostly out of the hands of everyday users appears during the manufacturing process. Errors can happen while the physical devices are being built. They can compromise the integrity from the start and make it much more difficult to secure the phone, tablet, or laptop after it is already in use.

Many different components go into every single computing device on the market today. More specific details about flaws in individual parts are covered below. However, putting all the different parts together also offers many opportunities to make mistakes or not follow stringent protocols. The best security protections need exceptional manufacturing practices for every single part and component.

Imagine if a tiny computer chip was manufactured incorrectly. Its function in the overall operation of the device is therefore weakened. Defects introduced during the process of putting everything together and installing firmware may also introduce vulnerabilities.

Another emerging challenge for manufacturers involves the increasing automation and digitalization of the product creation process itself. Computers are building computers. This means that the cybersecurity of factory must maintain even greater controls than ever before. If a malicious actor can sneak into the control systems of the manufacturing process, they may be able to affect and compromise every device that comes off the line.

Hardware Component Issues

Design flaws and manufacturing defects can affect individual hardware components just as easily as they can the entire computing device. Each of these needs to go through the same rigorous process that includes effective protection every step of the way. Things like processors, memory modules, and network interfaces are integral to the overall cybersecurity levels of the finished device. Any error or vulnerability in a single component affects the whole.

One of the unique challenges of securing the hardware components involves the supply chain itself. Usually, different parts come from different manufacturers and are shipped to the final production facility to be put together into the consumer product. Attackers may seek to interrupt the transportation and logistics process by tampering with the hardware components or even swapping them out for something with built in backdoors or malicious firmware. Physical security within the supply chain matters.

Firmware Weaknesses

Firmware is similar to software except that it is fully integrated into the computing device itself. These instructions tell the hardware how to function and what it should do based on the actions of the user. For the vast majority of devices, this comes preinstalled when you purchase a computer, laptop, phone, smart gadget, or other gear. The average user and even many IT team professionals cannot change this in any way by themselves. The manufacturer or firmware developer is responsible for any updates and bug fixes.

Any of the above-mentioned problems cause vulnerabilities that can be exploited by hackers. They may allow them to bypass security systems or any protections put in place with the onboard software. To stop this issue from affecting your personal life or business operations, it is important to take a proactive stance and address weaknesses before they become noticed by the wrong people.

Threat Detection for Computers and Mobile Devices

You already understand how important rigorous cybersecurity controls are throughout the entire lifecycle of any computer or mobile device. In order to protect everything from personal information to entire corporate networks, threat detection is of paramount importance in today's risky digital landscape. Everything runs on computers these days, and it takes a lot of effort to ensure their safety.

Businesses or individuals who buy desktops, laptops, smart phones, and IOT devices have virtually no control over design flaws, manufacturing issues, or hardware component oversight. It is up to the designers, developers, and factories to detect threats, overcome and block them, and ensure that every product they deliver is secure and will remain that way throughout its life span.

Not only do these efforts ensure that their customers are satisfied with the products, but they also bolster the brand's reputation and keep them competitive in the marketplace. From a buyer perspective, knowing that the creators took cybersecurity seriously gives a high degree of peace of mind. How is this possible?

Identify Risks and Weak Points

These essential tasks must take place during every stage of the computing devices' lifecycle.

- Designers should include threat assessments and risk analysis in the creation process so that they can choose the best hardware, systems, and set up overall.

- Developers must integrate their firmware and software choices into the hardware in effective and highly secure ways. They should evaluate the security of chosen components, protocols, and interfaces through a deep understanding of current and future potential risks.

- Manufacturers need to lock down the entire production process as well as ensure security throughout the supply chain. Any automation or digital processes they use should likewise be investigated for vulnerable spots.

- End users likewise need to stay vigilant for possible risks and signs that unauthorized access has occurred. Any errors or changes in the operations of the computing device may signify something wrong.

Design Security Solutions

Once risk factors are identified, it is the responsibility of each group or individual to create effective security strategies that protect design, development, manufacturing, and device use. For the first three categories, this includes building security principles into the architecture of the device itself through secure coding practices and thorough attention to detail.

Manufacturers must focus on high levels of quality assurance and attentive supply chain management practices. No one can leave cybersecurity to chance or assume that last year's practices are good enough to protect against next year's threats.

Organizations that provide computing devices to their employees are primarily responsible for designing security solutions for everyone in-house. However, the end user should also have sufficient training and education to understand risk factors, proper use of the hardware, security protocols, how to recognize attacks, and how to respond to minimize overall effect.

Implement Effective Protocols

Implementation of the well-designed security solutions is the next step. While many of these things could integrate with the automated systems used in the design, development, and manufacturing processes, it is still ultimately up to the people who control these things to ensure security. Education about risks, protections, and best practices should become part of an ongoing oversight strategy.

As with most other security issues, hardware protection needs constant effort. Regularly update firmware and software, configure the onboard security settings properly, remain vigilant against social engineering attacks, and prevent unauthorized physical access to the devices themselves.

Test and Monitor Continuously

Again, the need to test and monitor security protocols and potential weak points exists at every stage of the computing device lifecycle. Designers need to implement testing when they decide how to create the device. Developers also need to push hard and identify faults so that they can work as many as possible out of the finished product before it is even built. Manufacturers also need to conduct threat assessments and check their automated and manual practices for issues. This also holds true for supply chain management.

By implementing ongoing testing and monitoring practices, organizations take a proactive stance against current and emerging threats. The moment any vulnerabilities are identified in hardware components, firmware, and software configurations, they can take action before serious consequences occur. Common methods include vulnerability assessments, penetration testing, and security audits.

The specific schedule of how often to do these things depends on the computing device itself, the stage in its production or use cycle, and the general capabilities of the company or individual responsible for it. These are things that the organization needs to decide for themselves. A design team, for example, will test frequently to ensure that every stage in the process is as secure as possible. A company that purchases business laptops may only require monthly monitoring.

The time between tests can increase if there is automatic intrusion detection and prevention systems (IDPS) built into the hardware itself. This is one example of using tech to secure tech, which is a very common and highly effective way of bolstering overall cybersecurity efforts. These things also help to avoid human error issues or the potential issue of in-house malicious actors.

Best Practices for Endpoint Device Security

Endpoint devices are those used directly by people in their personal or professional lives. They encompass the entire range of computer gear from desktops at work, laptops in home offices, smart phones, tablets, IOT devices, gaming consoles, and more. If the ones you use are designed, developed, and manufactured with the most stringent security protocols in place, any cyber threats fall on your or your employer's shoulders. This is where things like app and network security come into play.

However, physically protecting the endpoint device from unauthorized access matters. Also, things like firmware updates and software patches must occur immediately in order to keep the overall security functioning. If you adopt these strict rules and develop security-focused habits, you can strengthen endpoint device security considerably. This helps protect the confidentiality of all types of digital assets.

Protection Against Unauthorized Access

- Implement strong password policies for device login
- Enable two-factor or multi-factor authentication
- Assign specific devices to unique individuals
- Disallow personal devices from accessing work networks

Use and Updates of Security Features

- Keep firmware, operating systems, and all software updated
- Install anti-malware and antivirus software
- Keep threat detection apps updated and active
- Enable robust firewall protection

User Education and Best Practices

- Train computer device users to follow all necessary safety protocols
- Stress the importance of ongoing security awareness education
- Create an easy-to-follow risk response process

Regular Endpoint Device Monitoring

- Set up regularly scheduled automatic testing and monitoring systems

- Backup data frequently while checking it for anomalies or unauthorized access

No matter who is using the computing device, cybersecurity takes ongoing effort to maximize its effectiveness at every turn. There are always new threats emerging, always new attackers and malicious organizations trying to gain control of the hardware, and always new challenges that designers, developers, manufacturers, and users need to be aware of and take proactive effort against.

Mobile Device Security Considerations

Smart phones and other devices you carry with you have become omnipresent for both private and commercial activities. These store as much sensitive information as larger computers that primarily stay indoors. You use a smart phone to access corporate networks, conduct financial transactions, and share personal data with others.

The same design protocols for designers, developers, and manufacturers still hold true. However, mobile devices are primarily under the control of the specific user after they purchase it or are assigned it by their company. Therefore, it is up to you to implement strong authentication mechanisms including biometric options like fingerprint scanning. It also helps to have a strong PIN as well as a unique name and password.

One of the specific security risks associated with phones and tablets is the common option to connect to public Wi-Fi. It is virtually impossible to know whether the network is secured properly or may be vulnerable to man in the middle attacks. Avoid using these whenever possible, especially when it comes to devices that access company information or networks.

Current trends that include the Internet of Things (IoT) devices, quantum computing, VR and augmented reality, and new connectivity options like 5G and beyond also add new concerns to the cybersecurity conversation. Emerging threats keep pace with innovative technologies.

Tomorrow's computing devices will need even more protection, monitoring, and strategic responses to keep businesses and people safe from malicious actors.

Every person in the modern world uses multiple computing devices every day. When you log on to your desktop computer at work, you expect to be able to handle important business tasks safely. When you relax with your laptop at home, you need to trust the built-in security features to protect your information and activities. When you tap and swipe on your smart phone, you should feel confident that the manufacturer did everything possible to make it safe for you to use.

The people who design and develop computer components and the final devices must integrate cybersecurity strategies and ongoing management practices at every stage of the process. This does not end once the product goes to market or gets shipped out to a company who intends to use it for business purposes. Monitoring, testing, firmware updates, and more are all an essential part of the process of making digital security truly bulletproof. All these things depend on the **Code** that developers use to create the apps and connect them to networks.

Code

Protect Your Programming: Secure the Code Throughout Its Development and Lifecycle

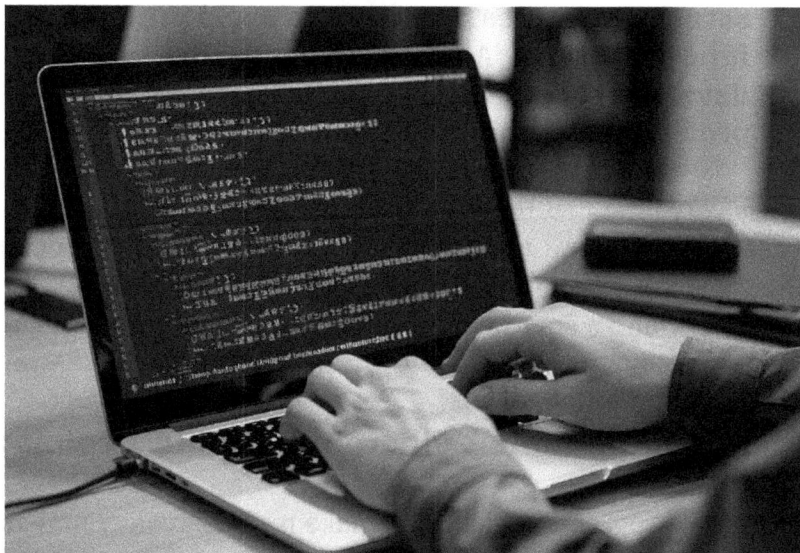

The second C in the PANCCD cybersecurity pyramid stands for "code." The process of coding or programming software that is used with a wide variety of computer devices needs just as much attention if you want to protect data, networks, and users. There are many stages in the development lifecycle of even the simplest app, and any potential weaknesses or exploit opportunities will be taken advantage of by malicious attackers of all kinds.

The world is increasingly reliant on software programs for everything. They control things as simple as shopping online for birthday presents to complex manufacturing plant production controls. This increased dependence on software comes with a heightened risk of cyber threats. Understanding the fundamental principles, processes, and procedures for protecting code helps to ensure confidentiality, functional reliability, and availability of programs you rely on every day.

Tip: Securing code is a multifaceted endeavor that first requires understanding of vulnerabilities, weaknesses, and exploits. Once a programmer recognizes those, they are better equipped to take security measures necessary to avoid present and future problems. All these things matter to developers, designers, IT teams, security professionals, and the companies who employ them.

There is very little that the average end user can do to increase code security. Your best bet is to only use software from reputable sources and keep programs updated. This is everyone's responsibility and a smart contribution to the entire cybersecurity world.

Understanding Code Vulnerabilities and Exploits

Professional programmers can still make mistakes. They may not be up to date on all the latest threats or ways to prevent them. The fact that cyber criminals come up with new technological methods all the time means that coders must do the same. It takes a lot of time, training, and experience to clearly identify risks associated with creating code for any purpose.

The following items represent a short list of common issues that software developers must be aware of and actively work to prevent.

Injection Attack Risks

This category covers a few different opportunities for cyber criminals to infiltrate the code. They could do this through SQL injection, cross-site scripting (XSS), command injection, and similar things. These actions are done after the program is released and the attacker sends on trustworthy data in an effort to manipulate how the application functions.

Broken Access Control

Any insecure access or authorization mechanisms make it easy for attackers to compromise the security of software, get inside, and do whatever they want with the information or assets found there.

Security Misconfiguration

At its most basic, a programmer who mis-configures the security code for a program leaves gaps in the protection and allows other things to sneak in All this is also something that a user may do after the app is released, the person or team developing the code must make sure it supports the users' actions if they want to prevent unauthorized access.

Cryptographic Failures

Besides using outdated or insufficient encryption protocols or algorithms, developers may accidentally leave hardcoded passwords in the program itself. These are the types of things that malicious actors can make use of. They scan for these types of vulnerabilities and pounce on them as soon as they are discovered.

Server-Side Request Forgery (SSRF)

Applications that are designed to fetch data or content from a third-party resource can be vulnerable if the attacker sneaks in the back door. Sometimes, cyber criminals look for attached resources as pathways to affect a main program. If the program fails to validate the resource's URL, the server-side request may gain access.

Buffer Overflow Vulnerabilities

These occur when a program writes more data to a buffer than it can reasonably hold. This may come about due to an error in the control programming and can lead to memory corruption, unauthorized access, and exploitation that allows for external code.

In the entire realm of code creation and software development, there are virtually endless errors and insufficiencies that can lead to cybersecurity problems. These are just a sample of the things that coders should be aware of when they start to create a new product. The huge number of possible vulnerabilities require constant vigilance, monitoring, and testing throughout the software development lifecycle.

Software Development Lifecycle: Secure Programming Practices

Cybersecurity has a lot of influence over many decisions that occur during the software development lifecycle. With a particular focus on safe coding practices, project teams need to do more than write clean code. As with every other layer of the PANCCD pyramid, security is a matter of starting out right, monitoring and testing frequently, and maintaining things throughout their useful lifespan.

These are the essential steps of the software development lifecycle:

Establish Goals, Scope, and Objectives

Before any programmer puts their fingers on the keyboard, extensive meetings take place with the project development team, executive groups, and other stakeholders to figure out what the program needs. What are the necessary features? What goals and objectives does this app need to fulfill? What is the scope of the project specifically in regard to budget and timeline?

Perhaps most importantly, everyone needs to be on board with understanding security requirements. There is no benefit in cutting corners to release things to market more quickly or save a few dollars.

Software Design Stage

The people actually responsible for creating the software now take over the project. They choose framework, create architecture, work toward design specifications, and make sure everything meets the aforementioned goals. Secure design principles must remain a top priority throughout. This includes things like access control, authentication protocols, and encryption.

Development and Implementation

Now is the time when the code actually gets created. Programmers with the most up-to-date and accurate knowledge, skills, and supporting experience transform the design into reality. Every line of code must follow certain standards and guidelines to minimize vulnerabilities and weaknesses.

A lot of code review, double checking, and oversight should be involved with this process so potential issues are caught as quickly as possible.

Function and Performance Testing

Not only does the finished program need to do everything the stakeholders want it to, but it also needs to function efficiently, reliably, and securely. Unit, integration, system, and security testing procedures are performed many times in order to limit problems. The specific types of tests performed depend on the purpose and scope of the software.

Software Deployment

After full testing and validation of the code and the overall function of the program itself, it is deployed to the end users. This is sometimes done in stages and may involve things like alpha and beta tests, which are quite common in entertainment and gaming industries.

For business-oriented software, especially that which is custom-made for a specific organization, deployment focuses more on configuring things to the established tech infrastructure and network. It also often involves integrating the program with in-house security protocols.

Maintenance and Updates

A developer team's job is not done after deploying a piece of software. The only way to keep things secure for the long term is to monitor emerging threats, notice new weaknesses, and send out patches and updates to counter them. This is one of the most important parts of the software development cycle from a cybersecurity standpoint. It is something that usually requires the cooperation of companies who purchased the software or individual end users.

No single piece of software lasts forever. The final stage in any program lifecycle is retirement. This is when the developers no longer support the program or offer any patches or updates. In some cases, it requires complete decommissioning of the software itself. This is especially true for bespoke corporate software. It allows for the complete removal of sensitive information and disconnection from any networks that may become vulnerable if the obsolete program remains.

The Software Bill of Materials (SBOM)

In the world of cybersecurity, new terminology related to software supply chain management has been in use and has evolved over the last few years. The Software Bill of Materials contains an inventory of all the elements that make up software components. These can include third-party or open-source components that are commonly used in software development. It also includes licensing and up-to-date security information.

The term itself mimics bills of materials used in manufacturing, which may list materials or parts of a physical product. For example, it would include the plastic case, glass screen, chips, screws, and wires, along with all the other components that go into a smart phone. A software bill of materials lists code assets and programming parts instead.

The SBOM and Cybersecurity

From a purely practical perspective, having a master list of software components with included information about licenses, potential threats, and risk assessments makes sense. This helps developers not only choose the right options for their projects, but also keep a record of already known vulnerabilities that they must compensate for or overcome. The SBOM for a particular project is therefore both a management help and offers assurance that everything possible is done for security's sake.

Deployment and Configuration Management

Creating code is not the only part of successfully completing a new piece of software. Deployment and configuration within its final use case is an important part of the process. There are numerous critical practices and considerations involving cybersecurity during this challenging time. As much as the client or consumers want quick and seamless delivery, they will appreciate thoroughness and the highest degree of safety and privacy possible more.

The following key principles, strategies, and techniques for securely deploying and configuring software applications greatly minimize security risks and enhance the overall safety of organizations and end users. While these things are not necessarily handled by the developers, they do play a vital role during this time. After all, security must be built into every app and program in ways that create a higher level of protection for every network it is on and every person who uses it.

Secure Deployment Practices

This first section includes many of the details described below. Before getting into the specifics, it is important to understand the value of creating a secure deployment practice strategy with distinct rules and protocols that the developers follow every time. The process should never include guesswork, going with your gut, or working from memory. Every individual may have a different idea about how deployment should work. Therefore, it is essential to override personal habits and create a step-by-step plan that covers everything.

To some degree, the client will need to be tweaked for different recipients, organizations, network types, and programs. Unless the development team frequently works on highly diverse code, it is possible to create a strategy that works for most of them. Creating this would be part of the project team's responsibility.

Baseline Configuration for Integrity and Security

One essential part of the overall deployment process involves creating a secure baseline image as a starting point and guide for ongoing software security. This clearly shows everyone involved is standardized and secure foundation including all necessary settings and configurations. A secure baseline image helps maintain consistency and compliance with whatever policies are necessary to satisfy stakeholder requirements or legal regulations.

These also help with future issues related to configuration drift. This term is used for the unintentional yet inevitable divergence from original configuration settings. In other words, software that is set up once properly will not remain that way forever. Things drift over time, whether through unauthorized or accidental changes, hardware or software upgrades, or faults in the code itself. A secure baseline image can help bring things back to where they are supposed to be.

Setting Access and Authentication Controls

Even the best and most secure code in the world will not stay safe forever if handed over to just anyone. While this is very different in the commercial software industry of PC games and phone apps, developers who create products for business use need to focus more on specific deployment for authorized users only.

This is frequently worked into the code itself through a system of role-based access controls (RBAC). These are either set up in the software from the start or configured so that the organization can do so themselves. Whichever way access and authentication gets handled, the process must be completely secure and inviolable. If developers are responsible for assigning specific user roles and granting access, they must work closely with the client to ensure proper functionality without vulnerabilities that cyber criminals can take advantage of.

Other elements of this part of the code creation process include integration of security features mentioned before such as multi factor authentication, biometric authentication, and built-in policies that enforce strong password creation and frequent changes. All these things must be explained during the deployment process.

Integrating With a Secure Network

Even the best written code and most securely created software in the world will still involve far too many vulnerabilities if it is not deployed to a secure network. This pertains only to programs used in corporate settings that control their own network in-house. Of course, there is no way for developers to have any control over a private user attaching a program to their own Wi-Fi or cloud servers.

In the deployment process, it simply makes sense to implement robust network security measures to minimize threats related to a piece of software. The code must allow for things like firewall use, intrusion detection and prevention systems, and more. Developers should work with whoever controls the network to make sure things integrate smoothly and securely.

Data Encryption and Privacy

Not only does any data related to the software operation need effective encryption, the code itself should be protected in a variety of ways. Developers must ensure that there are no backdoors into the inner workings of the program itself. While these types of breaches are less common than things that take advantage of network weaknesses or social engineering issues, for example, they are still threats to the overall cybersecurity realm.

Setting Up Patch Management

You already understand the importance of ongoing patches and updates to keep code secure for the long-term. Part of the configuration and deployment process involves setting up a schedule or expectations about when these patches are released and what users must do to maintain the program. New software vulnerabilities are discovered and taken advantage of all the time. This makes a coordinated patch management strategy and essential part of the entire process.

Developers need to create the patches and updates that deal with these issues, of course. Unless they also provide ongoing maintenance support, the actual updates are taken care of by the users. To make this as secure as possible, it is a good idea to code in as much automation as can reasonably be handled by the program. Built-in systems that scan for missing updates and send notifications are often a good idea.

Regulatory Requirements and Compliance

The world has rules for code, software development, configuration, and deployment. They come from a variety of international organizations specifically related to the tech world or unique industries like finance, healthcare, education, and government. If developers create programs associated with these or other regulatory bodies, they must ensure full compliance with all the rules they are legally bound to.

Some of these address the security and privacy of sensitive data directly. Others focus on financial and legal liability. Any programmers working in the medical sphere must remain aware of the Health Insurance Portability and Accountability Act (HIPAA), while finance industry developers must follow Payment Card Industry Data Security Standard (PCI DSS) rules. One common regulation that transcends sectors is the General Data Protection Regulation (GDPR) put forth by the European Union.

These and other compliance issues affect not only how code is created, but also the ongoing security controls after deployment. They go beyond quality programming practices to ensure effective encryption, stringent access controls, and data retention policies. If you do not meet the regulatory requirements, it could result in legal or financial trouble.

When Does the Developers' Responsibility End?

Every software development project comes to an end at some point. However, with the importance of frequent patches and updates for operational and security purposes, the project team cannot simply stop working after delivering the product. The answer to this question is complex and nuanced.

These days, developers need to recognize that ensuring code and software safety is an ongoing process that extends beyond delivery and deployment. It frequently involves collaboration with other stakeholders and the end user or client company itself. This can also include working closely with operations teams and in-house IT departments.

Although things like functional glitches and issues that affect user-friendliness or features that the customer wanted can affect the ongoing relationship, cybersecurity is most likely the main focus. It is always possible that weaknesses and vulnerabilities come to light after the software is deployed. People may use it differently than the developers intended or considered when they were creating the code. New threats may arise that challenge the integrity of the app.

To some degree, the original developers should be prepared to respond promptly to any news of security vulnerabilities no matter where they come from. Tracking the success of a software product is part of ensuring excellence and a high reputation for the developer team themselves. In the end, it does not matter whether this was a custom project for a specific company, or a retail product released to the consumer market. Feedback matters, and it guides the answer to this question about when developers are 100% done.

Besides being responsible for issues with the code of a specific piece of software itself, it also makes sense for developers to stay up to date with general information about emerging threats and security best practices. By participating in ongoing training and education programs, exploring security news in the greater tech world, and staying involved with the cybersecurity community helps increase individual and team value. This type of ongoing vigilance allows developers to take a proactive stance that serves both past app deliveries and future projects.

Eventually, most pieces of software are retired or dubbed obsolete and unsupported by the development team or the controlling company. This represents the final moment of the developers are responsible for the code they created in the first place. By this time, they are undoubtedly busy with the next greatest thing to release on the market.

Code is one of the broadest foundational parts of the entire PANCCD cybersecurity pyramid. It plays a vital and undeniable role in the practice of securing software applications against cyber threats. There are many aspects of code security that developers and other stakeholders in the process need to pay attention to. Understand what common vulnerabilities and exploits exist, establish smart development protocols and practices, and maintain a regular schedule of maintenance and updates. These basics allow organizations to take proactive measures to strengthen the final software products resilience against nefarious actors.

One of the most important takeaways from this chapter focuses on the importance of integrating security considerations every step of the software development lifecycle. Initial design and conversations with stakeholders need to include this information.

The programmers who create code must have the ability to follow the best practices and remain compliant with a diverse range of regulations. Finally, the end users need to understand their role in cyber threat reduction.

New code protocols, development frameworks, and cybersecurity risks appear all the time. Developers must stay up to date with the industry in order to provide the best service and deliver high quality, secure software. All these efforts help with the protection of the greatest asset: **data**.

Data

Secure the Data Vault: Ensure Protection of Your Most Valuable Asset

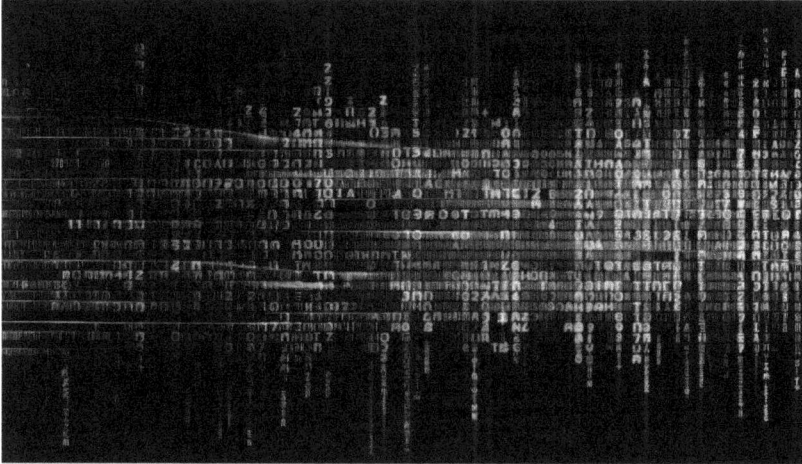

What is it that all the cyber criminals and hackers want? Although "control" or "power" or reasonable answers in the metaphoric sense, what these nefarious actors really want is data.

Whether they specifically want personal information like Social Security numbers, financial details like credit card numbers, or data related to business dealings, medical records, or other sensitive topics, their main goal is to access data without being stopped or noticed.

This same data is the lifeblood of businesses, government and private sector organizations, and individual lives. In today's digital age, every person has a high amount of personal data floating around on the Internet and in business systems. Companies control even more and rely on it for everything from day-to-day operations to future success.

Tip: As cyber criminals grow in number and capability, the overall security threat level increases. Individuals and automated systems constantly seek out vulnerabilities or ways to exploit insecure networks, systems, or devices for their own personal gain. Protecting this data has never been more important.

It is not enough to create strategies to deal with issues after they occur. Of course, any data protection plan should include a strict series of steps to take as soon as a breach gets noticed. Outcomes improve when organizations adopt a proactive approach to safeguarding information against potential threats. Explore the myriad ways in which hackers, criminals, and espionage actors target data and gain access to it. Then, learn how to create and implement effective data protection strategies to mitigate the risk. This not only protects you on a personal level but also the stability and reputation of any organization.

Understanding Threats to Your Data

As the main target of all cyber-criminal behavior, data faces myriad risks and challenges that organizations and individuals need to learn to protect against. The entire process of safeguarding information includes every other part of this PANCCD pyramid. In today's world of massive interconnectivity and digitalization, both human and automated threats are all trying to get at important, personal, and potentially lucrative sources of data.

These things range from highly sophisticated and malicious systems and that operate on a large scale and have a lot of equipment and resources behind them to simple and accidental breaches caused by a flaw in an app's code, network security, or due to user error. Understanding these threats makes it much easier to resist them. It also helps companies create effective protection strategies. Not only can you prevent data loss and manipulation, but you can also learn what to do if such an issue occurs.

First, it helps to know what cyber criminals do with the data they steal. Why do they want access to all of it in the first place? Is data specifically the main goal of every malicious attack?

Secondly, how do these threats manifest? In other words, how do hackers and other bad actors gain access to the data they want? Exploring the answer to these questions starts with the basics and more specific information described below. However, different organizations need to look at their own strategies, protocols, and operation styles in order to find faults or weaknesses. It is impossible to suggest a blanket protection policy that will cover every single company.

That is the end goal of all cybersecurity efforts. By gaining insight into various threats facing your data, you can prioritize protection efforts, implement more proactive measures, and take quick and decisive action if anything does go wrong. These efforts can mean the difference between a successful and safe business and serious reputation, financial, and legal ramifications.

What Kind of Data Do Cyber Criminals Want?

Although an individual hacker or criminal group may have their own specific targets for the type of data that they can hold for ransom, sell to someone else, or use for their own purposes, these are the most common types of data that these malicious actors want to get their hands on.

Personal Identification Information (PII) – This type of data includes names, addresses, phone numbers, Social Security numbers, tax ID numbers, and any other details that specifically identify an individual person. These are often used for identity theft and fraud situations.

Intellectual Property (IP) – The realm of intellectual property includes many different things. Some common data that a criminal may want to get their hands on includes trade secrets, proprietary information about products, things like publications or copyrighted creations, and other things not released to the general public. These are either used for corporate espionage, to gain a competitive advantage in a specific industry, or to be sold on the black market.

Financial Data – This is one of the most common types of data stolen by hackers. Things like bank account details, credit card numbers, and financial transaction records have an obvious value to people who want to improve their own financial situation.

Authentication Credentials – If a cyber-criminal can gain access to usernames, passwords, security tokens, and other authentication information from individuals or company systems, they can then get at many other types of data. This allows them to sneak past security protocols and take whatever nefarious actions they desire against the accounts, systems, and networks.

Online Credentials – On a more personal rather than corporate perspective, bad actors attempt to get access to login credentials for web-based platforms and applications. This can include things like email accounts, social media pages, and even e-commerce sites. Once they have these details, it becomes easier to engage in identity theft, phishing attacks, or other social engineering efforts.

Corporate Details – Criminals specifically interested in their own corporate gain or data-based ransom will attempt to gain access to or steal corporate data. This can include things like business plans, confidential communication, client lists, and other proprietary information.

Health Records – PHI, or protected health information, includes medical history, diagnoses, prescribed medication records, treatment plans, and other personal details about an individual.

These things can help criminals perpetrate identity theft, blackmail against the person, or insurance fraud among other things.

Sensitive Personal Information – Data targeted for more personal and nefarious reasons includes sensitive information about an individual that can be used for blackmail, extortion, and public embarrassment. This can include anything from private messages or emails to photographs, videos, or even web histories or product and service purchase lists.

Although cyber-criminals may target an exceptionally wide and diverse range of data types, the reasons to do so are usually financial in nature.

What Do Cyber Criminals Do with the Data They Steal?

In the grand scheme of criminal behavior, data thieves and hackers share the same goal as most other people who engage in illegal activity: financial gain. Instead of robbing a jewelry store, holding up a bank, or committing fraud, they use their technological know-how and equipment to obtain unauthorized access to both individuals and businesses to steal sensitive data.

Cyber criminals then exploited this stolen data in a variety of ways. Identity theft is a huge and lucrative issue. For example, if a hacker gets personal information like Social Security numbers and financial details from multiple people, they can gain massive amounts of credit very quickly. Other data theft reasons include more advanced concepts like espionage and extortion. Some demand ransom money for holding data hostage from the rightful owners.

Some data theft occurs for more personal reasons. People with vendettas against another individual or those who want to destroy someone's reputation can easily do so if they have access to private details. This is a much less common form of cyber-crime, but it may be an issue especially for people in prominent positions or with a lot of wealth.

As you can see, any data breach or security compromise puts not only personal or professional information at risk, but it can also cost businesses a lot of money to recover or fix. Protecting data is the base layer of not only the PANCCD cybersecurity pyramid, but also the foundation of all protection efforts in any setting.

How Do They Gain Access?

Explore the methods and techniques used by criminals to breach security measures and gain unauthorized access to your most valuable asset: data. If you can understand how they infiltrate systems and networks, you are better equipped to create an effective defense strategy.

Social Engineering – Manipulating the human element in your data protection plan is one of the most common and effective ways of gaining access to the information they want. This can include things like phishing emails or text messages that trick people into clicking a nefarious or fake link, or even fraudulent phone calls asking for specific details.

Malware and Ransomware -- If the cyber-criminal can inject malware or other malicious software into a computer system or network, they can use it to sneak into databases or other data storage options and access whatever they want. Ransomware automatically encrypts data and demands payment for decryption. This may be done for financial gain but is also a method for gaining access to the data itself.

Man-in-the-Middle Attacks – These involve the interception of communication between two parties or in-house announcements. The attacker can essentially eavesdrop on any exchange and thus access, steal, or modify the data related to the transfer.

Other Data Breaches – Every single part of an organization's digital operations can introduce vulnerabilities to the entire thing. Hackers and malicious actors know how to look for these and take advantage of them. Data breaches can occur whenever unauthorized access happens. This is why the whole realm of cybersecurity must include a robust and multi-faceted strategy.

Data Protection Strategies

All the above-mentioned threats would cause serious consequences for the organizations or individuals that experience them. To protect data, your most valuable asset, implementing both proactive and reactive policies can help safeguard protected information from a wide range of criminal behavior. Cyber threats continue to evolve and grow ever more frequent and sophisticated. The following strategies and best practices can help you maintain privacy, integrity, and availability of data while avoiding serious financial and legal repercussions.

There are a huge number of different things a company can do to protect their data. Many of them are covered in other parts of this book such as encryption, strong network controls, and training employees against social engineering attacks. Creating a comprehensive strategy and making sure everyone is on board is one of the best defenses against unauthorized access, manipulation, and theft.

In the end, it does not matter if you represent a large corporation with unique trade secrets, a healthcare clinic with private patient details, or a small e-commerce store that handles customer credit card numbers. You may need different data protection strategies, but all the essential elements must be in place to fight against the rising tide of digital criminal behavior.

What Data Assets Do You Need to Protect?

Assessing your assets involves conducting a comprehensive evaluation of the different types of data that are stored, used, and transferred through your systems. It is important to know what you have if you want to protect it effectively. Look into things like health sensitive the data is, the perceived value and why someone would want to get their hands on it, and if it falls under any regulatory requirements. Some types of information may need special protection if the risk is greater or if it is involved in things like finance or healthcare.

Figuring out how much the data is worth is not something that you can quantify exactly. Instead, this process has more to do with understanding what cyber-criminals will want to get their hands on more. The riskier it is, the more protection you should put in place. However, that does not mean you should skimp on the security for other information.

Identify Possible Threats and Risks

What are the potential sources of danger that could compromise the privacy, integrity, or access to the data that your organization needs? Finding the answer to this question begins with understanding the fundamentals of cyber-crime. This book on cybersecurity Simplified clearly describes many of the potential weaknesses, vulnerabilities, and threats related to all types of data breaches and network access.

When creating your specific strategy to protect data, look into the types of things that are commonly perpetrated in your industry. Also, keep a record of past direct attacks against your organization. This can help you create a more proactive plan to deal with similar things in the future. However, it is important to look beyond the basics or past experiences when formulating your plan. Always be on the lookout for emerging risks and new methods for protection.

Establish Data Protection Goals

Strategic planning of any kind always includes the establishment of goals and objectives. Be specific. Of course, your ultimate goal is to protect all your data from cyber-criminal behavior and accidental breaches. This is not detailed enough to help you create protocols to do so effectively.

Look at the specific threats and the particular types of data to determine what goals you should make for each one. For example, minimizing the risk of data breaches requires unique goals related to things like network security, data encryption, and employee training against social engineering efforts.

On a grand scale, one of the most important objectives of all cyber protection strategies involves maintaining continuity of business operations. Not only do you need your company to keep operating smoothly for your own financial purposes, but it is the only way to establish client or customer trust and grow successfully for the future.

Implement Digital Security Strategy

After you establish goals for protection against a diverse range of threats, the time has come to put the strategy into practice. This is not a simple process in most cases.

There are many different avenues of protection that an organization must follow for the strongest threat elimination possible.

The strategy must be both precise and holistic in scope:

- Create a roadmap of key objectives and milestones to achieve maximum security
- Allocate resources including talent, tech, and money to the new plans

- Upgrade existing infrastructure and introduce new tech and equipment where necessary
- Install new encryption models, authorization access systems, and more
- Establish oversight and control processes to make sure everything goes smoothly

One of the most important parts of implementing a new data protection plan is making sure everyone knows their role and can conduct themselves in ways that support the strategy. The pinnacle of the PANCCD pyramid is, after all, people.

Train People with Data Access

Just as data access is the goal for cyber-criminals, people are the first line of both defense and weakness in the fight against it. Training and awareness programs are crucial parts of a comprehensive protection strategy. They give employees at every level of the organization the knowledge and skills necessary to recognize and mitigate risk. It is important to include ongoing education in the overall data protection policy.

What types of things should training programs cover?

- Basic cybersecurity awareness of risks and threats
- Organizational issues that may arise if a data breach occurs (Business shut down or loss of employment)
- Safe communication and operation rules (data handling, email safety, etc.)
- More secure username and password usage
- Information about appropriate response to issues

Include Regular Strategy Reviews

Even the best security strategy will not work optimally forever. If the criminals' efforts change and get better all the time, it stands to reason that your proactive plans and responses must also improve. The only way to make these things effective is to review the strategy frequently and update it when necessary.

Schedule periodic evaluations of the data protection measures in place. Stay informed about emerging threats related to new techniques or technologies appearing in your industry and the world of cybersecurity at large. Take a systematic approach to check your processes for efficacy and adequacy. Make sure they can mitigate risks today and tomorrow.

These strategy reviews must involve people on every level of the organization. You need to have C-suite executives involved in the process, but it will primarily fall to the IT department, managers, and project leads to handle everything. This is also a great opportunity to brush up on regular employee training to make sure they do not forget all the things they learned.

Privacy and Regulatory Compliance

Data needs protection from the strongest cybersecurity strategies and in-house protocols that your organization can put into place. However, amid all the efforts you put forth, it is also important to understand that different types of sensitive data are also controlled by various regulatory organizations on both the national and international levels. These can affect how you secure data, especially when it comes to individual privacy.

Understanding Privacy Regulations

Ignorance of the law is no excuse for breaking it, and there are considerable legal consequences for failure to secure data appropriately. You must navigate a complex web of privacy regulations and rules that are created to control the collection, use, and disclosure of all different types of personal information. These regulations are designed to help make sure you have the right cybersecurity protection in place.

One example is the General Data Protection Regulation (GDPR) created by the European Union. Since 2018, the EU has put this collection of privacy regulations into effect. It covers everything from the rights of individuals to control their own information to various guidelines and permissions for organizations who handle and use data.

In the United States, the California Consumer Privacy Act (CCPA) is representative of the strictest requirements having to do with data protection, transparency, and rights. Other states seek to comply with these rules to make sure that everything is sufficiently covered.

If your organization has anything to do with medical records or healthcare information, you must also understand the Health Insurance Portability and Accountability Act (HIPAA), which defines how you can and cannot share personal medical information.

Other data protection laws exist. Canada has a Personal Information Protection and Electronic Documents Act. There is a US federal law called COPPA that governs online collection of children's personal data. The financial industry ruling bodies have a few required regulations of their own for things like credit card information. The LGPD in Brazil outlines many of the same things as the GDPR. It is up to you to know which laws and regulations you must legally follow in the area in which you do business.

Why Does Regulatory Compliance Matter?

These and other specific rules about personal data handling do more than promote effective cybersecurity strategies. Since they are either laws or legal rules and regulations, you must follow them or face legal and financial consequences. For example, large corporations who fail to uphold GDPR rules could be fined up to €20 million or 4% of international annual profits.

Create a Strong Incident Response Plan

Identifying threats, shoring up weaknesses, securing systems, and complying with various regulations or all interval part of protecting your greatest digital asset: data. Although preparation and ongoing updates and reviews are essential, they are still not foolproof when it comes to preventing data breaches, access, and theft. *No cybersecurity strategy is ever 100% effective.*

This is why one of the most important parts of overall operations includes the creation and deployment of a strong incident response plan. Therefore, if or when something goes wrong, *and it will*, every employee knows exactly what to do to fix things as quickly as possible.

This requires strategic planning and ongoing training alongside a culture of transparency that welcomes input from every level of the organization. In other words, if a part-time administrative assistant sees something questionable, they should feel that they are welcome to tell someone just as much as a top executive would be.

What goes into a quality incident response plan?

Understand the Importance of Having One

Benjamin Franklin famously said, "If you fail to plan, you are planning to fail." This holds true for almost everything in life, but especially something as important as cybersecurity and data protection. If you do not have an incident response plan in place before a breach or other criminal activity occurs, you and your employees will have no idea what to do in order to stop it and return to normal operations as quickly as possible.

The risks are considerable for any size business or organization.

- Additional loss of data access and control

- Prolonged downtime in which you cannot operate safely
- Financial losses due to downtime and other factors
- Regulatory fines and legal consequences
- Damage to the brand or individual reputation

An incident response plan provides a systematic approach for identifying data breaches in the first place, analyzing what is going on, containing the issue, and recovering from it. These plans also help give stakeholders peace of mind that you are taking their private data seriously.

Establish an Incident Response Team

In the People section of this book and the PANCCD pyramid, you came to understand that people are both one of the highest risk factors and layers of protection when it comes to the entire realm of cybersecurity. With that in mind, understand that everyone needs to be equipped with proper training, education, and focus when it comes to incident response.

The establishment of a specific incident response team streamlines the process. Create one to include people from various departments and disciplines within the organization. Do not rely on the IT department alone, because they are not the ones necessarily integrated with all day today activities that could trigger or be affected by threats. Instead, build the team from IT, cybersecurity, the legal department, communications, and the executive leadership. Choose individuals for their diverse skills, willingness to work together, and proactive nature. Try to avoid including people who would be more interested in keeping problems secret rather than solving them efficiently.

After choosing the team, defined clear roles, responsibilities, and escalation procedures to make sure all their efforts are coordinated effectively and that communication lines stay open at all times. These people should have additional specialized training on a regular basis, especially when strategies and protocols get updated or when the tech securing data changes.

Identify the Incident Type

When something goes wrong, the response must be immediate and comprehensive. The first step is to identify what exactly happened. The response team should identify the nature and scope of the problem, determine the potential impact on business functions, and categorize it as one of the predefined varieties of security issues.

Incident types may include malware infections, phishing or other social engineering attacks, denial-of-service attacks, unauthorized access attempts, data breaches, or many other things specific to your organization's systems were networks. Define these ahead of time while educating the incident response team. While things may occasionally happen outside the scope of expected behaviors, it helps to have a set plan in place for the most common before they occur.

How can an organization quickly identify what type of cybersecurity incident took place? If they have to do specifically with a person, such as malicious insider breaches or an employee clicking on a fake link in a phishing email, categorization becomes much easier.

Outside of someone specifically telling the team what happened, most organizations rely on various detection mechanisms. These include things like intrusion detection systems (IDS) and security information and event management solutions (SIEM), tools that monitor networks, and authorized user activity logs. The response team must have quick access to these things and know where to look for the information they need.

Communication and Notification

Hiding problems does not make them go away. When it comes to data protection, it is vital that an incident response plan focuses on maintaining transparency, accountability, and trust during security incidents. The response team must stay in contact with each other in order to fulfill all their duties as efficiently as possible. Is likewise important to keep any stakeholders informed about what is happening and what you are doing to mitigate damages. Another thing to consider is communicating with regulatory authorities and law enforcement agencies if necessary.

Set up these protocols before incidents happen. Establish clear channels that everyone is comfortable using so there are no delays or other issues during a data breach or threat. Clear and timely communication helps to minimize confusion, reduce public relations issues due to the spread of misinformation, and foster a greater sense of collaboration and shared responsibility.

Be thorough when defining the communication and notification systems and rules for these types of problems. Some of the essential procedures include establishing designated points of contact, defining escalation paths, and outlining the content and format of communication messages. You may want to create incident response templates that everyone needs to use in order to streamline the process and reduce extraneous information.

The more you plan and organize before recognizing a threat to your data, the easier it will be to shore up the defenses and manage the outcome. By following predefined procedures, every aspect of the incident response process becomes simpler and faster. This is the best way to demonstrate accountability, commitment to addressing the problem effectively, transparency, and reliability to everyone involved and the public.

Effective Response Procedures

The well-trained incident response team knows how to communicate quickly and clearly to stop the various actors from gaining further access to data. Besides opening these channels, you must establish what the effective response procedures are. After a data breach occurs is not the time to figure out what to do about it. Instead, create a structured framework for analyzing, containing, eradicating, and recovering from cybersecurity events.

Detection is always the first step in the entire process. Then, communication begins to inform important stakeholders and the response team. Each member should then dive into performing their specialized tasks to stop the incident or start the recovery process.

Initial Assessment – What is the nature and severity of the incident? Where and how did the breach occur and what data was compromised?

Containment – The team must stop unauthorized access and make sure that all other threatened data is now safe and secure. This step may include network isolation, disabling compromised accounts, or adding stricter security rules to different systems or apps.
Eradication – Get to the root of the problem and get rid of it. This step involves many potential actions depending on the nature of the incident. Some involve malware removal, system reconfiguration, data recovery steps, or similar actions.

Recovery Stage – How will the organization recover from the incident? This is sometimes a drawn-out process that involves everything from getting business operations back online to creating and deploying a public relations marketing strategy. The incident must be analyzed carefully through forensic investigations. It might be a good idea to add things you learn into the incident response strategy.

Ongoing Testing and Training

As with every other part of the cybersecurity pyramid, protecting data and having effective incident responses depends a lot on frequent maintenance of all systems and protocols. Test to make sure that the plan will work if something goes wrong again. Keep the incident response team up to date on training for new threats. Always remain on the lookout for emerging weaknesses or vulnerabilities. Refine your response capabilities and keep those lines of communication open.

Cyber criminals hunger for data for a wide variety of reasons. It is the greatest bait for their malicious activity and your greatest asset at the same time. As the foundation of the PANCCD pyramid, it is what all other security efforts, threats, strategies, and vulnerabilities rest on.

Every individual who uses the Internet or a connected device needs to understand the best and safest methods. Every business or organization must have more stringent plans in place to identify, stop, and recover from incidents.

When you prioritize data protection, implement robust protection measures, and foster a culture of cybersecurity awareness and resilience, you are better able to mitigate risks, safeguard sensitive data, and maintain the trust and confidence of stakeholders and consumers. The world of cyber-crime changes so quickly these days. Proactive measures and diligent adherence to the best practices are necessary to navigate the challenges and overcome the threats to your data.

The PANCCD Pyramid for Cybersecurity Simplified

The six layers of the PANCCD pyramid cover the intricacies of protecting digital assets and fortifying defenses against an array of cyber threats. The whole data protection realm includes a diverse range of factors.

The people, apps, networking, computing devices, code, and data represented in the acronym all play essential roles in understanding and safeguarding digital infrastructure. The whole thing needs a holistic approach that also focuses on minute details that cyber criminals are ready to take advantage of for exploiting.

As we near the conclusion of this book's exploration into cybersecurity, it should be evident that achieving the kind of protection you or your organization needs requires more than setting up a few rules or implementing reactive measures. Planning ahead, establishing effective protocols, and maintaining a proactive stance and commitment are necessary parts of staying ahead of the next breach.

> **Tip: The PANCCD pyramid serves as a guiding framework to remind everyone not to focus on individual parts of this type of challenge. They are all interconnected and dependent on each other. By prioritizing cybersecurity education, training, and awareness, you can better prepare the people involved for taking care of all the other elements. By strengthening application security, securing computers and devices, fortifying network defenses, and imbuing the entire coding and development process with security, you can build a truly resilient system that will be able to withstand the relentless onslaught of threats.**

Your journey to maximum cybersecurity does not end here. As specified in many areas of this book, the world of digital criminal behavior always changes and creates new ways to get at the valuable data that hackers and other malicious actors want. Embrace emerging technologies, stay ahead of industry trends, and foster a culture of innovation and collaboration. You must stay as forward-thinking and proactive as the criminals themselves if you want to stop them.

The Tight Budget Basics for PANCCD

Cyber security takes effort, a lot of personnel power, and carefully designed tech systems. These all cost money, so it makes sense that an organization's security strategy needs a budget. Many people may believe that robust protection starts with expenses. However, in the ever-evolving world of cyber threats, resources are finite, and it makes sense to look for solutions that are truly tight budget. It is essential to create foundational principles of effective cyber security that cost nothing and still contribute to overall improvements.

Organizations and individuals must understand what threats look like at every level of the PANCCD pyramid. From identifying risks and vulnerabilities to implementing access controls and avoiding social engineering methods, the basics remain unconfined by budgetary limits. These are the cornerstones upon which resilient cyber defenses are built.

If you are in any way responsible for a business's overall cyber security plan, you must cultivate a strategic focus on practical approaches and cost-effective solutions. In the sections below, you can discover how to navigate the complexities of the digital protection world without breaking the bank. No matter how limited your resources are, it is possible to establish a stronger shield against hackers, malware, and other nefarious apps or actors seeking access to or control of sensitive data.

Foundational Principles of Tight Budget PANCCD

The PANCCD pyramid of cyber security represents each element that must be protected from digital threats. They all represent both the source of vulnerabilities and the opportunities to create a more bulletproof shield against assets and systems. You may question if it is possible to shore up the defenses without a big budget. While the answer to that question contains many complexities and depends largely on the organization in question and their digital operations, there are many options to explore without the need for a significant budget.

By embracing the foundational principles, it is possible to lay a solid foundation for tight budget cyber resilience and protection against nefarious actions. This is an essential part of operations especially for startups and small to medium-sized companies that do not have extra money to invest.

Understanding the Holistic Nature of the PANCCD Pyramid

Earlier sections of this book described each layer of the PANCCD pyramid in detail. These include:

- People
- Apps
- Networks
- Computing Devices
- Code
- Data

While in-depth knowledge about each level is a necessary part of creating an overall strategy, it also leads to the reasonable assumption that effective measures take money. As tight budget as anyone would want their protection efforts to be, cost and resource allotment is an integral part of any organizational improvement.

However, when you take the time to understand the holistic nature of this cyber security pyramid, you will be better equipped to create a strategy that does not rely on monetary investment. Instead, it creates a culture of security and emphasizes practical approaches that imbue day-to-day operations with safe practices from start to finish.

Cyber protection does not lie solely on the shoulders of IT departments and expensive software that automatically scans for intrusions, handles alerts, or even locks things down on its own. Every facet of an organization, from its people to its processes, plays a role in the overall security strategy. It is possible to overcome financial constraints with this type of holistic understanding and preparation.

Building a Strong Defense without Breaking the Bank

Many recommendations for security seem to indicate that you need a bottomless budget when it comes to protecting against digital threats. The tight budget approach to PANCCD challenges this notion with a collection of methods for fortifying your position without reallocating funds from other important things. It is important to take a pragmatic approach and keep that holistic overview in mind. This is not a matter of simply ticking off boxes on a to do list. Instead, these things work together to empower you toward a more secure status.

At its core, building a strong defense requires a shift in mindset. You must learn to prioritize resourcefulness, creativity, and flexibility. At the same time, you must encourage these habits and capabilities in employees at every level of company operations. Expensive security tools and technologies can make things easier. However, there are always alternatives.

Take Stock of Your Assets and Resources

Before you make any decisions or changes to regular operations, collect information about what you already have that can help your overall cyber security efforts. No organization is without some kind of protection from the start. This introspective process involves evaluating tangible assets like hardware and security systems and less obvious ones like knowledge, expertise, and dedication to the overall protection of business assets like data.

Just like you would create a financial budget, create an asset and resource budget by conducting a comprehensive inventory of what is already available. This will help you gain a clear understanding of your strengths and weaknesses so that you can make informed decisions about the changes that will improve things over time.

Several parts of this process exist. It helps to begin by identifying the assets you need to protect most of all. This frequently focuses on the bottom layer of the PANCCD pyramid: data. While exploring potential access routes hackers may take advantage of, also figure out what the impact of such a breach would be. This is all part of the risk assessment process.

Then, move on to identifying the most useful security features of existing systems. Hopefully, you are already using these to the greatest effect. These can include things like firewalls, built-in malware blockers, secure password creation protocols, and similar things. Also take the time to look into any third-party or cloud-based connections. What advantages do they provide for your protection?

It is impossible to improve things if you do not know the current status of your organization's cyber security capabilities. After identifying assets and resources and related risks, it is time to make changes within the bounds of a tight budget system.

Leverage Open-Source Solutions

While it is very difficult to depend on free resources only for a comprehensive security system, there are ways to leverage the power of open-source solutions. Open-source software offers a cost-effective alternative to proprietary tools and technologies. There are many respected options created, updated, and maintained by knowledgeable experts. Most of these are highly accessible and flexible for unique organizations and system types. When you tap into the collaborative efforts of the global community, you can discover tools, frameworks, and libraries that eliminate or reduce weaknesses.

One of the most prominent advantages of using these tight budget options comes from their transparency and adaptability. Unlike paying a high price for custom software, your in-house team can scrutinize and verify the security of the app itself. This utilizes the asset of knowledge and skills that you already have within your company. Transparency increases trust because there are frequently multiple experts working to improve the solution all the time. It also helps that open-source solutions often have robust communities where you can find help.

Very few companies stay the same size forever. This focus on growth often runs parallel to and need for additional investment or higher budgetary limits. Open-source solutions offer better scalability, so they can benefit your overall cyber security plan now and into the future. Best of all, they do this without increasing costs in any way.

Create Security-focused Processes

The idea of creating a culture of bulletproof security has been mentioned several times already. This is especially important when attempting to remove the level of protection from a direct association with finance. A strategy that includes processes that make sense will fortify the defenses effectively without requiring additional investment. Company culture is a part of this because employees from the part-time admin to the executive C suite will understand their role in the overall process. By weaving security into every process and strategy, it improves overall consciousness and therefore mitigates risks.

This depends largely on establishing clear policies, procedures, and guidelines that govern not only things like security protocols and responses to threats but also how all projects and tasks are managed from the start. Integration is the key word. Every member of the team or department should have a defined role and responsibility to play in protecting everything that goes into success. This includes data handling, security assessment maintenance, process auditing, and the tiniest day-to-day tasks.

Creating these processes is especially important in the app development and coding part of the PANCCD pyramid. No matter what type of digital product you work on, the entire lifecycle must include security baked into the process. This is the power of proactive process creation, which transcends any possible benefit of reactive solutions once the threat is detected or a weakness discovered.

Prioritize Security Awareness and Training

Some onboarding and training processes cost a lot of money. Large corporations outsource education and come up with fancy presentations or documentation that all employees must use in order to keep their job or get a certain security clearance. While these things can help and make sense for organizations with considerable budgets, it is possible to prioritize awareness and training in a tight budget security situation.

People are at the top of the PANCCD pyramid. They represent both the weakest link and the first line of defense in the whole realm of cyber security. It simply makes sense to boost awareness and training on how to recognize weaknesses, actively take steps to prevent them, and respond to threats if they happen.

How do you do this without spending extra money? When you have a comprehensive knowledge of every level of the security pyramid, you can create in-house training materials that cost nothing more than a bit of time and energy. If you want printed materials, you may have to invest in extra ink or toner for the office printer.

There are many ways to make this happen. Awareness and training initiatives encompass a wide range of activities that can be as simple as emails or handouts or as complex as multi-session classes. Some important topics to cover include:

- Raising awareness about phishing techniques and other social engineering issues
- Training on effective password security with tips on choosing safe ones
- Safe software, Internet, and network use tips
- Guidance on identifying breaches and responding to them effectively

You can find many more pieces of specific advice in the earlier chapter on People and putting training and education efforts into action. Most importantly, promote open communication and encourage people to get involved with the entire security process. Never penalize any employee for speaking up if you want a workforce that is vigilant, resilient, and well-equipped to defend against cyber threats.

Strategic Investments: Prioritizing Protection

Many of the same things mentioned above that a company may be able to do without increased expenditure can benefit from additional resource allotment. Attempting to create a truly bulletproof cyber security system without any investment at all can lead to dangerous gaps in protection. There are always hackers ready to take advantage of such vulnerabilities. Organizations without a budget for these things carry a greater risk than those who can strategically invest in their security strategies.

What are the most important things to focus on when prioritizing protection? As with all other aspects of security, the answer to this question depends on industry, operations, network types, software used, the people in your organization, and many other factors. Some types of companies are simply not as attractive to cyber criminals as others. Therefore, the following list includes things to consider within the bounds of your specific situation.

- Security software and tools
- Employee training and awareness programs
- Secured network infrastructure and equipment
- Professional security assessments and audits
- Expert help for regulatory compliance
- Secure cloud services and solutions
- Insurance coverage for security incident recovery

Why Budget Matters to Resilient Cybersecurity

Ultimately, aiming at completely tight budget cybersecurity may not make much sense for most companies. It is important to have that critical discussion on the relationship between finances and the effectiveness of protection methods. Of course, ample monetary resources help you implement sophisticated security technologies, use top-notch training systems, and hire talented specialists.

Budget plays a nuanced role in shaping your overall strategy. Bigger does not always equal better. Therefore, it is important to explore how you can maximize the impact of your investments even with limited financial resources.

Can You Create a Bulletproof Security System for Free?

Most people will automatically answer "no" to this question. The idea of creating a company-wide, resilient cybersecurity plan without investing money seems unbelievable. Cyber threats change, grow, and evolve all the time. You need to stay up to date with the best practices and latest technology if you want to stop hackers and other nefarious actors from gaining access to systems, networks, and data. It is rather unrealistic to expect a completely foolproof solution without any budget, but you might be surprised how small and effective one could be.

Most of the tips outlined in this chapter focus on the most budget-friendly options available today. Open-source tools and resources combined with the attached communities' help can provide many of the digital resources you need: tools, libraries, and frameworks. Free security resources and educational materials available online can train employees in the best methods.

Finally, tight budget cyber security relies on an in-depth understanding of the PANCCD pyramid and all the best practices outlined elsewhere in this book. It takes resourcefulness, creativity, agility, and strategic planning to maximize efforts at every level. Now that you understand this, continue to dive into the specifics that will further strengthen the shield that will protect you from current and future threats.

The Risk Formula Mapped To PANCCD

To maximize both security levels and affordability, is important to effectively assess risk before diving into any purchases or new protocols. Understanding and managing different levels of risk lies at the core of building effective defense strategies. Various threats exist at every level of the PANCCD pyramid. To have effective organizational security, you must recognize each one and take actions against it.

Risk assessment is one of the most important parts of the whole process. By implementing targeted mitigation measures, companies can reduce the likelihood and impact of security incidents. This will then protect networks, systems, and data from the greatest number of threats. Of course, as you learned in the chapters on the PANCCD levels, threats come in a wide variety of specific types and focuses. Using a risk formula to compare them with existing vulnerabilities is both a highly complex process and an essential one.

What Is the Cyber Security Risk Formula?

Consider this simple risk formula:

Risk = Likelihood x Impact

While it is the simplest explanation for how to calculate cyber security risk, it suffers from that simplicity at the same time. There are innumerable factors that go into both the likelihood and impact of a specific problem that may arise within an organization.

Likelihood – This refers to the probability of a cybersecurity event such as a direct threat, measured vulnerability, exposure, or failed or weak security controls.

- High – Motivated cyber criminals with great capability facing many weaknesses or vulnerabilities

- Medium – Capable threat sources facing stronger controls and protections

- Low – Weak threats going up against stronger security measures

Impact – While some equate this to the severity of a threat or incident, that is another amorphous word that does not truly define what you need to know to conduct a proper risk assessment. What would be the actual outcome and effect on the organization?

- High – Will result in tangible or financial loss and impede business operations while damaging reputation

- Medium – Limited loss of resources or money and harm to the organization's mission, reputation, or progress

- Low – Some financial or asset loss with interruption of operations and minor reputational damage

Some mapping formulas use a graph that lists severity from low, medium, high, and critical and likelihood from impossible to probable to possible. The terminology does not specifically matter. The meaning behind the categories remains the same.

Different specific risk formula models exist in the IT world. For example, there is a manual approach called the FAIR Framework that focuses on monetary risk, loss event frequency, and the financial loss magnitude. More organizations turn to AI and ML systems to conduct cyber risk quantification based on automatically gathered or inputted data.

An important part of the formula that should be handled first involves specifying an acceptable level of risk. It is impossible to eradicate all cybersecurity vulnerabilities and threats completely. Therefore, an organization should categorize them into avoid, control, accept, and monitor segments. This helps you focus on the highest risk or highest impact issues, which will do a better job of mitigating negative outcomes.

No matter what specific formula or method you choose, there is no denying that risk assessment and measurement are important parts of planning and implementing the overall cyber security strategy. They help you decide where to allocate resources, how much money to spend, what type of training to provide for employees, and more.

The PANCCD Pyramid and Cyber Risk Formula

Every level of the pyramid should have its own risk formula assessment completed to get a comprehensive picture of the organization's security standing. This section outlines reasons to do so, methods to make it possible, and other details that you can use to create your own calculations. However, every single company is different, and it is virtually impossible to come up with a single plan that will work for everyone.

Things like the number of employees, different departments, in-house or cloud networks used, implemented apps and systems, and more all affect the outcome. The simplest equation mentioned above is nowhere near sufficient for the complexity of the entire cyber risk assessment process. Above all else, be thorough. Do not skimp on using the best, latest, and most complete data. Use secure tools to analyze it. Choose the best risk model for your specific circumstances and needs.

Take the time to review the earlier chapters on the PANCCD pyramid for more in-depth information about each layer. Now, learn how to use a risk formula mapped to these specific factors to maximize overall protection.

People

As the top layer of the cyber security pyramid, people represent both vulnerability and protection in an organization's overall strategy. As the most dynamic component of the entire thing, it is also one of the most difficult to fit into the risk formula. Human behavior is much more unpredictable than things like firewalls or code.

Most of the protection factors that come from people directly depend on their awareness and ongoing training and education. Consider how people affect both the likelihood and impact of any cyber threat.

Risk Likelihood

How well trained are the people in your organization to recognize social engineering, mitigate risk, and respond to breaches quickly and effectively? The likelihood of damage from a hacker or nefarious bot depends on the answers to these questions.

Human error, negligence, lack of comfort with technology, and susceptibility to social engineering attacks all increase the likelihood of successful criminal activity. Thus, the risk factor increases.

Well-trained, security-conscious, and supported employees who understand how and are welcome to communicate with higher ups if something occurs can act as a vital layer of defense. This decreases the likelihood of attacks.

Risk Impact

The level of risk associated with people either making mistakes or not following protocols properly depends largely on what systems or data they have access to and how many other security measures are in place. Make determinations for specific individuals or departments in order to figure out the right proactive measures to take.

When you integrate strong and accessible security awareness programs, provide ongoing training, and foster a culture of vigilance and accountability, you will be able to reduce the likelihood and potential impact of any risks associated with people in your organization.

Apps

The software programs you choose to use for business have a great impact on the risk assessment and response practices you must implement for effective control. The cyber risk formula applies to apps the same way it does for anything else. They represent a critical component of your digital infrastructure and day-to-day operations. However, they also introduce vulnerabilities that hackers and other bad actors can exploit.

Risk Likelihood

How accessible is the app to criminals who want to use it as a doorway into your databases or other assets? Most connect to the Internet these days or integrate with third-party services. These things increase the likelihood of successful attacks. Complex and intricate pieces of software may also have more built-in vulnerabilities to consider.

A lot of the responsibility of reducing threat likelihood falls on the developers' shoulders. However, if you source apps from outside your own company, it is up to you to determine if the source is sufficiently focused on protection.

Besides the actual app itself and its integration with your systems, it is also important to stay on top of the overall threat landscape and cyber security news. As new threats emerge, criminals may go after specific types of apps more often. This may increase the risk likelihood from medium to severe. Awareness of these changes can help you take proactive steps for protection.

Risk Impact

Apps often allow for more access to data or user information that cyber criminals can exploit or use for their own nefarious purposes. The potential impact to your organization depends on the type of software, what it is connected to, and other factors.

For example, a data breach that gives a hacker access to all user logon information can expose proprietary data or intellectual property in ways that can lead to financial losses and even legal trouble. Service disruption through a DDoS attack on the program will have a less serious impact as it will only disrupt operations for a short period of time.

Before choosing an app to use for your organization, conduct a thorough risk assessment to identify potential weaknesses both with the program itself and in conjunction with your systems and practices. Avoid outsourcing to teams or using apps from brands that do not have the same dedication to security as you do. Also, monitoring and updating is an important part of reducing both the likelihood and possible impact of a breach.

Networks

The next layer of the PANCCD cyber security pyramid involves network infrastructure. In order to identify potential threats and assess the likelihood and impact, it helps to understand the capabilities and techniques used by these criminals. More in-depth information about this is covered in the earlier chapter. When focusing on the cyber threat risk formula alone, the same challenges present themselves as in the other categories.

Every organization is different and has different networking systems and practices. In general, however, these threats may have an overall greater impact because networks connect everything. Once a hacker or other criminal gains access to this interconnected highway, they have greater opportunities to access systems, data, and more.

Risk Likelihood

How do you determine if the likelihood of a network breach is high, medium, or low? It all starts when you choose on-site or cloud-based networking in the first place. Each comes with their own vulnerabilities and levels of potential control. Risk factors include things like misconfigurations, outdated software, insecure protocols, and weaknesses related to user error.

To test and reduce the likelihood of cyber-criminal access, keep things updated on a day-to-day basis and conduct comprehensive vulnerability assessments and penetration testing frequently. How often your organization does these things depends on current industry threat levels, emerging cyber methods, and other factors.

Risk Impact

As mentioned above, the impact of network breaches is frequently more extreme than issues that stem from things like phishing emails or denial of service attacks. If a criminal gains access, they can severely disrupt or shut down operations completely. Unfettered access to data, intellectual property, and business systems can cause massive financial loss, reputational damage, and cause trouble with regulatory groups.

These types of far-reaching consequences affect the integrity, availability, and confidentiality of many different types of assets. Within the cyber security pyramid, this is one of the more severe and potentially damaging levels when assessed with the risk formula model. Therefore, a considerable part of your organization's overall security strategy may have to focus on network protections.

Computing Devices

The potential for risk at this level may be quite small if your organization does not use a lot of physical computers or devices for regular operations. Of course, these days, every single business uses physical tech to manage operations, communicate with clients or customers, and do other essential tasks.

Both physical and digital protections must remain in place in order to reduce overall risk and make recovery practices effective. When using a risk formula to assess these things, the outcomes will be very different for every organization. Also, this layer of the PANCCD pyramid is highly associated with people and the vulnerabilities and production factors they bring to the table.

Risk Likelihood

The more devices, the higher the risk. With a greater variety of computing devices, the risk increases even more. For example, a small business who only uses desktop computers to handle operations will have a much lower likelihood of cyber criminals gaining access to these devices. A larger organization with desktops, laptops, company smart phones, tablets, and IOT devices has a much greater likelihood that a bad actor will gain physical access to the device or be able to access them digitally.

Another part of the likelihood side of the risk model involves evaluating the effectiveness of security controls that protect the devices themselves. This includes things like complex access or log in information, encryption methods, patch management strategies, and more.

Risk Impact

If the cyber-criminal gets a hold of a single computing device because someone forgot their company smart phone on the bus, they will not have much opportunity to make a big impact on your business. If a massive data breach occurs due to an unnoticed vulnerability shared across all desktops in the IT department, the impact is much greater and more difficult to recover from.

Most specifically, you will not be able to use affected computing devices for a potentially long time. Returning to regular operations after such a disruption can involve high costs and a great need to appease stakeholders and the public who rely on your brand security.

Protecting the computing devices used by your organization depends on both physical control of the items themselves and digital protections to prevent external access. The risk formula that looks at likelihood and impact must take many factors into consideration that are specific to your situation and use practices.

Code

Although the entire world of programming is quite complex and affects many different parts of your company's operations, it is also the one layer of the security pyramid that is most straightforward when it comes to the threat formula. In short, you would not hire a team to develop code if they did not have a firm grasp and focus on current security measures or ongoing dedication to patches and security updates.

Risk Likelihood

The likelihood of any risks related to code depends largely on choosing the right developers from the start. Things like frequent code changes and updates reduce the likelihood of a severe breach. External threats emerge all the time, of course, and the people responsible for the code must keep up with these changes. Assessing risk likelihood in this realm is one of the simpler tasks, although it is by no means easy.

Risk Impact

It may also be easier to assess the potential impact of a cyber breach when it comes to code. Each part of the program used to power some part of your organization has a particular purpose. If it becomes vulnerable or a malicious actor finds a way to sneak in or affect it, you should be able to identify the outcome with relative surety.

The impact of code vulnerabilities varies depending on the critical nature of the affected application, the sensitivity of the data it provides access to, and how far-reaching the access is. The impact of a security flaw in a financial application is understandably much more serious than a simple bit of code that automates holiday messages to the employees.

Data

The foundation of the PANCCD pyramid – data – is the ultimate goal of every criminal actor and malicious automated threat that works within the cyber security realm. Ultimately, the assessment of risk for every other part of the pyramid comes back to data at its core. This is what the criminals want access to or control over.

Therefore, it should be the main focus of the risk formula equation. Understanding the likelihood and impact of data breaches will help you determine where to prioritize efforts to safeguard it from unauthorized access, theft, or manipulation.

Risk Likelihood

How likely is it that criminals can access your organization's protected data? The answer to this question depends on innumerable factors related to the other levels of this pyramid. It also takes into account the effectiveness of security controls and measures currently in place, the frequency and sophistication of cyber at, the level of awareness and preparedness of your employees, and uncontrollable external factors.

Analyzing the likelihood of data vulnerabilities, weaknesses and systems that protect it, and probable or possible threats is a necessary part of constructing an overall security strategy that works.

Risk Impact

When deciding on that strategy, take the time to weigh the potential impact of each type of data breach issue. The consequences of security incidents on the integrity, confidentiality, and availability of these assets can impact your organization in every way. You may run afoul of regulatory compliance and have to pay fines or other recompense. You will lose reputation with stakeholders and consumers who trust their data with your company. Some of the greatest risks have to do with operational disruptions. A severe data breach can stop your organization in its tracks and make it very difficult to recover.

The cyber security risk formula looks at the likelihood and impact of many types of breaches, vulnerabilities, attacks, and other damaging situations. In order to create the most effective, efficient, and secure cyber strategy, you must assess these factors for every layer of the PANCCD pyramid as it pertains specifically to your organization.

Time-based Cyber Security Formulas and Basics

No matter what cybersecurity issue arises, speed matters for every aspect of the detection and recovery process. The ability to rapidly analyze, respond, mitigate effects, and return to normalcy greatly reduces the overall impact of the attack. This improves outcomes for the organization itself, customers and clients, and other stakeholders. It is a huge part of maintaining a positive reputation in your industry or sector.

There are four basic parts of time-based cybersecurity to consider.

- Time to Analyze
- Time to Respond
- Time to Mitigate
- Time to Normalcy

Tip: To maintain an effective cyber security system, the total response time from the organization should be shorter than the time when the adversary breaks in and has access to any

systems or data. Speed is of the essence when it comes to an effective threat response.

This chapter sheds light on how companies can optimize their incident response processes, reduce downtime, and improve resilience against all levels of threats. The world of digital criminal activity changes all the time, and any plan or strategy must remain flexible to keep response times low. Learn how to set realistic time-based goals and equip all levels of the organization with the knowledge and tools necessary to safeguard systems and assets.

Time to Analyze the Risk and Impact

The first step in minimizing response time for any security incident involves analyzing the risk and its overall impact on operations and the brand itself. This can occur before an attack even happens and should be part of the overall cyber protection policy. However, if something does go wrong, it makes sense to figure out what it is as quickly as possible.

Swift and accurate risk analysis allows the company to determine the severity of the incident and make swift and smart decisions about an appropriate and effective response. When you can understand the nature and scope of the nefarious act as soon as possible, it makes every other part of the response more streamlined. The organization knows how to allocate resources, mobilize response teams, and put mitigation strategies into place.

Not only do response teams need to analyze the risk associated with a specific issue, weakness, or action, they must also estimate the impact of something going wrong. At the very least, this helps them understand the appropriate responses. It also helps maintain clear communication with executives and stakeholders who may want to downplay issues for public relations. Managing expectations and keeping things real is an important part of protecting the brand's reputation. Ignoring or taking too long to analyze the impact can lead to a more severe outcome.

How can organizations reduce their time to analyze risks and impacts at every level of the PANCCD cyber security pyramid?

People – Analyzing Human Factors and Behavioral Risks

This is one of the most important and difficult parts of the entire cyber security protection system. It is very difficult to analyze what weaknesses people may introduce or what behaviors they will perform and when. In order to reduce the time it takes to do so, focus on the establishment of clear protocols and response procedures as well as sufficient training. These things will automatically reduce analysis time.

Shortening the analysis of impact requires a system of advanced analytics and monitoring tools that can identify nefarious or risky behavior as soon as it happens. Digitalizing human-related activities helps organizations make the analysis process much more efficient.

Apps – Fast Figuring for Software Vulnerabilities

Once an organization has chosen apps to work with, they should already have plenty of information about risk factors and the impact of a breach. Onboard and automated scanning and testing tools that continuously assess software vulnerabilities are the best way to reduce the time to analysis if a threat occurs. These are the types of things that the cyber security team must monitor continuously in order to handle risk and impact factors for applications smoothly.

Networks – Up-to-date Analysis of Infrastructure Risks

The steps involved with reducing time to analysis for networks aligns very closely with those for apps. Use security intelligence feeds, stay abreast of emerging threats, and automated scanning and testing tools to make analysis faster. It also helps if you segment networks while setting them up or expanding them. This proactive approach allows the IT team to quickly see exactly where the problem is so they can assess its impact and take action.

Computing Devices – Evaluate Hardware and Firmware Quickly

Deploy endpoint detection and response solutions to constantly monitor for threats and incidents. Automate patching and updates as well as device configuration. Build behavioral tracking into the computing devices to reduce the time for noticing an issue in conjunction with the person who is responsible for using the gadget itself. The more knowledge you have about what can happen and what actually happened, the faster you can assess its impact.

Code – Speed Comes from Continuous Oversight

In-house or outsourced cyber security teams may have no or limited programming knowledge. They are not responsible for checking the code to analyze risk while it is being created or used. Therefore, the primary ways to speed up the time to analysis depend on built-in code analysis tools and secure review processes. Again, the holistic integration of security testing throughout the development workflow is essential to minimize the time to analysis of risk and impact if anything nefarious should occur.

Data – Reduce the Time to Analyze Impact

As the primary target of cyber criminals, data issues have the largest impact on the entire security system for an organization. Reducing time for analysis primarily focuses on figuring out what the bad actor accessed, if they changed anything, and what the company needs to do to recover from it. It is also important to rapidly assess how a data breach will affect stakeholders and customers or the general public. This brings every department, including marketing and public relations, into the mix as soon as possible.

In general, the time to analyze the risk and impact of any issue depends on built-in processes and tools that provide real-time information. This is just one part of time-based cyber security, however. In order to minimize overall impact, the time to respond and stop the attack is even more important.

Time to Respond – Stop the Attack

Identifying and analyzing attacks is the first essential step in being able to stop them. The world of cyber-criminal activity is relentless and ever-changing. Threats loom at every turn, and your organization must be able to counter any incidents immediately in order to mitigate damage. The time to respond is an essential metric that must always be the focus of improvement efforts. You only have so long before the hacker or bot does immeasurable or severe damage that your company cannot recover from.

Explore fundamental principles and strategies that organizations can take to improve response capabilities and counter cyber threats as quickly as possible. Detection is only the first step. You need to have a strong incident response plan in place that everyone understands. A well-coordinated and timely response can mean the difference between containment and escalation. This takes not only knowledge and dedication from related teams and individuals, but it also takes the right tools and systems already deployed to neutralize the threat before they can inflict irreparable damage.

Identification and Classification of the Attack

It is impossible to stop an attack if you cannot identify exactly what it is and what type of response it needs. Different types of access or infiltration need different responses. No matter what, however, it needs to be fast. The ability to swiftly recognize and classify security incidents makes it possible for you to take immediate action to minimize potential damage. A lot of this depends on the automated detection mechanisms put into place well before any threats occur. It also relies on establishing clear protocols for incident reporting and escalation. These points will be explored in more depth in later paragraphs.

How do you speed up the identification and classification process? Digital monitoring systems play a crucial role, computing devices, and applications for signs of any unauthorized access or suspicious activity. They can even catch the outcome of successful social engineering efforts if configured correctly. Choose systems that collect sufficient data and analyze it based on robust threat intelligence guidelines. It is simply impossible to do manually what the automated options can do. Today, AI and machine learning play a huge role in the identification and classification of cyber-attacks.

The overall efficacy of the systems still depends on the information that powers their decision-making process. Either the original developers or the tech team for the company needs to establish clear protocols that define the threats themselves. What is a security incident? What are the different classifications one can fall into? These and other decisions make it possible to recognize them quickly when they happen.

Alert Triage and Prioritization

Once an individual or, more likely, and automated system identifies a threat or breach, the next step is to determine how risky it is and to prioritize where it falls on the overall response plan. Organizations with complex tech systems and multiple endpoints probably have a huge number of alerts and notifications coming in on a regular basis.

Which ones should you pay attention to first? While part of this triage process is handled by the programs themselves, it also falls on the shoulders of the cyber security people to speed up the process. Come up with a functioning strategy to prioritize these alerts based on severity, relevance, and potential impact to the business. Knowing what to respond to immediately and then doing so changes the outcome from disaster to something manageable.

The fastest way to handle this part of the process involves a proactive system of defining objective criteria for alert evaluation. Keep things consistent and know exactly what factors to consider when it comes to prioritizing responses. Alert triage frequently focuses on possible or probable outcomes for the organization as a whole. Things that could lead to more financial or reputational loss get higher priority. Attacks that put the company at risk of regulatory noncompliance should likewise be handled swiftly.

Activation of the Rapid Response Team

You have identified the risk and attack, categorized it by type, impact, and importance, and now need to respond as quickly as possible. As mentioned in earlier sections of this book, resilient cybersecurity requires the creation of a rapid response team. These individuals should know exactly what their role is in any response process, how to communicate with each other to avoid delays and mistakes and have set protocols in place for what to do and when.

Swift and coordinated action makes all the difference. The rapid response team could either be integrated into the overall threat alert system or be brought in only once the triage process is complete. This depends on various factors, so it is important for you to determine what method works best for your company. The team typically consists of security professionals with diverse skill sets. Therefore, they may be part of the identification and prioritization process too.

Automated alerts and instant communication that puts the response team into action are the quickest ways to make this happen. Once the incident is identified, team leaders or coordinators immediately know that they have to mobilize everyone else to stop and contain the attack. The primary responsibilities at this point include opening lines of communication and establishing methods of coordination such as virtual "war rooms" or physical meetings to create an effective plan.

The best way to make this go as quickly as possible is to ensure that every member of the response team knows their role precisely. You pick these people for their skills, knowledge, and capabilities in things like communication and collaboration. A dedicated team of experts can respond to the incident quickly enough to minimize the impact and restore normal operations.

Fast Containment Procedures to Minimize Damage

The final step in the Time to Respond portion involves stopping the attack and sealing it off from accessing other devices, networks, and data. Preventing further damage is one of the most important responsibilities of the rapid response team. The sooner this happens, the less damage is done.

It is not enough to identify the threat itself. You also must identify the affected systems and assets. Throwing up walls in back of an invading virus, for example, does little to stop its spread. Think proactively and explore the furthest reaches of potential access to take targeted action to contain the threat. Some of these containment procedures may include restricting user access, isolating affected devices, and even shutting down segments of the network.

Once the cyber-criminal activity is stopped and contained, the next time-sensitive task is to mitigate the damage and the scope of the attackers' affect.

Time to Mitigate the Damage

Stopping the threat and containing it as quickly as possible is incredibly important. However, it is not enough if you want your organization to avoid serious consequences down the road. The time to mitigate the damage is the third step in the time-based cyber security formula. Every second continues to count as your rapid response team now turns its attention to fixing the problem.

Retroactive efforts matter but taking a proactive approach to all cyber security issues is the best way to make everything happen quickly when a threat emerges, or a breach happens. The deployment of rapid response capabilities takes over where all your earlier efforts fail. This is where clear protocols and the human element of a highly automated system come into play. With the response team's necessary resources and authority, the mitigation process goes much more smoothly.

The overall company culture of vigilance and resilience also matters at this moment.

Rapid Damage Control Strategies

Everyone, from C-suite executives to the admin assistant who fell for a phishing attack, needs to know their role. The implementation of rapid strategies helps everyone mitigate the impact of security incidents more easily. For some, the instructions should include an order to stop doing anything and immediately shut down their computing devices or log off of the network. For others who are more involved with the damage mitigation process, it takes a clear step-by-step plan to overcome problems introduced by the attacker.

Containment has already occurred by this time That is part of the *Time to Respond* part of the equation. Now, threat neutralization is the main goal. This can include deploying patches and updates that the IT development team creates or perfects quickly, removing malicious software or unauthorized access points from compromised systems, and resetting logon credentials or access privileges.

Cyber Security Incident Recovery Efforts

The threat is removed, but the damage has been done. Recovery efforts now turning to removing the threat entirely and shoring up protections so does not happen again. This is a point in the process that requires quick action for the later process of returning to normalcy. However, you must also proceed with a high degree of care and attention to the best security practices.

The recovery often involves reinstalling software, restoring data from backups, and boosting the security measures. The response team or IT department continues to assess the impact on business operations and communicates with people who are responsible for things like regulatory compliance, public relations, and more.

Collecting Threat Data for Later Analysis

While not particularly time-sensitive in an immediate fashion, every single cyber security event that occurs within an organization must include full data collection procedures. The more you know about the attack and what worked to mitigate it, the better equipped you will be to deal with issues in the future. As with every other business activity from product development to marketing and beyond, data collection and analysis is an essential part of overall success.

Time to Normalcy – Returning to Operations

The cyber threat is identified, stopped, and contained. The next focus for time-based recovery involves how long it takes to return to normal, everyday operations. Some data breaches and compromised vulnerabilities may shut down everything. Others are, by definition, short-lived, but the company needs to stop working in order to make sure the threat is eliminated completely. This critical phase of incident response involves the organization striving to restore normalcy to ensure service continuity, minimize downtime, and keep stakeholders and consumers happy. The following steps or strategies will help you expedite this process to mitigate the overall impact.

Restoration of Critical Systems

If certain systems are turned off, you need to turn them back on again to use them. Prioritize the recovery of essential IT infrastructure, applications, affected computing devices, network segments, and services. This must be done in a controlled manner that double checks everything before continuing with normal business tasks.

Depending on the severity and nature of the incident, it may be necessary to implement temporary workarounds or alternative solutions to maintain operations to some degree. Getting some systems back online as soon as possible while long-term remediation efforts are underway can help alleviate some of the problems associated with downtime.

Data and Asset Recovery

If the response team or other departments have not yet identified the affected data and assets damaged by the cyber security incident, this must be done as soon as possible. Quickly work to recover and restore them to the safest state that existed before the incident occurred. Things to pay attention to include financial information, intellectual property, operational data, customer records, and similar things.

This step in the process makes it slow down because it is sometimes difficult to determine the extent of data loss or the corruption of existing information. For example, it may be difficult to check a customer information database that was corrupted against past information that may not have the latest accurate details. It may be necessary to establish longer-term recovery priorities for less essential data. Security and accuracy trumps speed in most cases.

Verification of Network Integrity

Finally, before returning to full normalcy, take extra time to verify that everything is secure and functioning normally. Check network and system integrity to make sure it is free from lingering threats or vulnerabilities. This may not occur before full resumption of normal operations does. However, it is an important step in shoring up defenses against future problems. The organization must weigh the cost vs. benefit of completing this validation step before any regular operations or putting it off until some normalcy is restored.

Resumption of Normal Operations

The longer it takes to resume normal organizational activity, the more money and reputation it will lose. Businesses need to function to grow and succeed. Not only does any disruption get in the way of day-to-day tasks, client or customer communication, and development projects, it also plays havoc with house stakeholders and the general public view you. Getting up to speed again on a limited basis is better than keeping everything shut down until the rapid response team gives the 'all clear.' Just make sure the network segments, systems, and computing devices you do turn back on have no cyber security issues lingering in the shadows.

Speed is of the essence at every stage of the cyber security response process. Proactive protocols and the creation of a skilled rapid response team provide innumerable benefits. However, the ability to quickly analyze the threat, stop it in its tracks, and recover will do more for the long-term operational success and reputation of any organization.

How do you convey these essential truths in an effective way that gets the attention of people who can actually implement change? There is no doubt that the PANCCD pyramid of cyber security provides a highly effective understanding of the entire concept. Now is the time to communicate it effectively to the people in charge who can make positive change happen.

Boards of Directors: Communicating the Foundational Basics of PANCCD

Any large organization needs to have the Board of Directors on board with cyber security changes and improvements. These people who are elected by shareholders to oversee the management of the business are ultimately responsible for creating policies, choosing executives, strategizing, and ensuring optimal function and regulatory compliance. They are involved in everything from assessing risks to financial reporting to public relations and more.

Tip: With your in-depth understanding of the PANCCD cyber security pyramid, it makes sense to consider how you can communicate these levels and best practices to the Board of Directors. The ability to do this helps ensure that the company has necessary resources, strategies, and oversight in place to mitigate risks effectively. Of course, any Board would want this.

As a cyber security professional, consider these methods of helping executives and shareholder representatives understand how best to protect the organization from the ever-evolving world of digital threats.

The Cyber Security Landscape – The Threat Is Real

The most important and sometimes difficult first step to communicating effectively with a Board of Directors involves supporting the idea that cyber security threats are real and omnipresent. You can introduce this concept at a specific meeting, but it helps to keep the members abreast of any emerging issues or high-profile news stories that involve data breaches or similar outcomes.

Comprehensive Overview of Evolving Threats

Cover the answers to these questions:

- What pertinent threats exist that can affect your company?
- What evolving industry trends can introduce new vulnerabilities to your system?
- Are there any pertinent regulatory developments that require new efforts?
- What have other companies dealt with that would create serious consequences for your company?

Point Out the Dangers of Failure to Prepare

Perhaps one of the most effective ways to effectively communicate cyber security issues is to clearly outline what would happen if these threats were allowed to take hold. The board members should understand the digital dependencies that make the organization run and thrive. These are people deeply involved in protocol creation, strategizing, and making smart decisions. Especially point out regulatory risks, potential financial losses, and damage to brand reputation if appropriate and proactive steps are ignored or delayed.

Risk Assessment, Management, and Preparedness

Introducing the framework of PANCCD to the Board of Directors needs to go further than describing what the levels mean, related risks and vulnerabilities, and potential outcomes. Risk assessment, threat management, and preparedness must take the holistic nature of the pyramid into account. This helps the members of the board gain valuable insight into the cyber risk posture across all levels. Combining the specifics with an overall view makes decision-making, resource allocation, and resilience a natural part of business.

PANCCD Protocols and Strategies

People – Encourage the board members to look beyond the executive levels and outward toward stakeholders and consumers. People use the technology, introduce weaknesses, but also play an important role in defense. Communicate the need for ongoing training and a culture of support.

Apps – Underscore the pivotal function of software for organizational success and the importance of choosing or developing the best options for security. Highlight the significance of an automatic or enforced update and patching protocol to counter vulnerabilities and exploits.

Networks – Educate board members on the foundational importance of these digital highways that facilitate the flow of information, communication, and allow for regular operational function of every single thing the company does. This omnipresent and interconnected element of the entire cyber security pyramid deserves prioritized investment.

Computing Devices – Many people take their smart phone or business computer for granted. Members of the Board may be as laissez-faire as device security as the average worker. Introduce the idea of mitigating risks by physically and digitally securing not only what is stored within but also the devices themselves.

Code – Organizations with developmental teams that create bespoke apps or systems for their own purposes need to focus on secure coding practices and rigorous reviews. Even those that outsource or use open-source or commercial products must understand the essential nature of quality, integrity, and frequent updates.

Data – Make sure members know that this is the ultimate goal for all cyber criminals: access to or control of data. This foundation for the PANCCD pyramid encompasses everything from the organization's intellectual property to regulatory controlled consumer information.

Investment and Resource Allocation

As mentioned above, many Boards of Directors strive to find ways to save money and cut costs to increase profits. As the cyber security professional tasked with communicating essential truths to them, you must enforce the idea that is suitable budget matters. It helps to outline the potential financial devastation that could occur due to a cyber-attack, data breach, or regulatory noncompliance that can result in serious fines. Even the smallest interruption in operations can lead to operational downtime that cuts into revenue considerably.

Following the levels of the PANCCD cyber security pyramid, these are the primary investment needs of any organization that wants to minimize risk.

- Training and education programs for employees at every level

- Secure and up-to-date apps with the latest automated protection and detection systems such as firewalls, intrusion detection, and event management apps

- Access control tech for physical network rooms and top-quality cloud-based network protections

- Endpoint protection solutions and access control to computing devices

- Talent acquisition for developmental projects or investment in outsourcing to high-quality programmers

- Data protection through onboard monitoring systems that offer real-time, digital oversight for every aspect of operations

Ways to Save Money Without Skimping on Protection

In truth, most Board of Directors members and other people responsible for organizational budgets want to spend as little money as possible. If you cannot effectively communicate the importance of investing in a robust cyber security system, the company will not have the protection it needs.

To soften the blow of discussions about resource allocation, keep information about ways to save money ready. Refer to the chapter on Tight Budget Cyber Security for some ideas.

Communicating effectively with any organization's Board of Directors is much simpler if you, as a cyber security professional, claim a seat at the table.

The next chapter covers how this can happen and how to make the collaboration process simpler and more effective for the overall implementation of threat protection strategies.

Serving on a Board as a Cybersecurity Professional

As more executives, shareholders, and other Board of Directors members become increasingly savvy about the rise of cyber threats, they also become more open to the idea of adding a specialized security professional to their ranks. The days of old school leaders who ignore tech risks and the possibility of catastrophic outcomes are disappearing. If you are a highly talented and successful cyber security expert, it is quite possible that you will receive an invitation to become an interval part of an organization's board.

New developments in the security industry as a whole have caught the attention of governments and regulatory bodies. The Securities and Exchange Commission, for example, is pursuing an option to require cyber security expertise and oversight disclosure for Boards of Directors. Protection has simply become an integral part of operational and financial existence and success.

Desired Attributes of a Board Cyber Security Expert

Once the decision to invite a cybersecurity professional to a board of Directors is made, the selection process begins. It is difficult to pinpoint the exact attributes that every company will look for. They all have different specific needs, existing skills within the board, and levels of interest and tolerance for new ideas. Tech sector companies are obviously more open to feeling empty seats with individuals skilled in IT, software engineering, data analytics, and similar specialties.

Technical Knowledge and Security Expertise

It may seem obvious to state that a cyber security professional must have a strong educational background in technology, information systems, innovative computers and devices, and similar things. Chief information security and technology officers have STEM degrees well beyond a simple bachelors.

Organizations look for top talents who align their expertise with the specific industry or niche that the company fills.

A Holistic View of a Company's Needs

The increasing prioritization of cyber security requires plenty of business experience that helps the new board member communicate effectively and get things done. No one is going to hire a simple IT specialist to fill this role. Instead, they will look to existing company executive leadership teams to find the right person. They must have the ability to stand back and look at the big picture of how to combine security protocols, systems, and improvements with the non-technical things that executives and shareholders care about most.

Experience and Proof of Success

No Board of Directors will invite a cyber security expert on their knowledge and talent alone. Proof of successful breach detection, prevention, and recovery in the past will go a long way to convincing leaders of your right to sit on the board. Individuals with desired background experience have often held roles as a Chief Information Security Officer (CISO) or the CEO of a tech industry corporation.

The right board cyber security candidate can come from any sector and hold any specific title. If you have the above-mentioned qualities and experience, you can have a say in how a company protects itself, its shareholders, and consumers from the dangers of cyber-criminal attacks.

Integrating with the World of Corporate Governance

Unless the Board of Directors is extremely open and ready for comprehensive overhauls of the entire cyber security system in place, do not expect to make sweeping changes after joining. It takes time to integrate fully into a specific company's governance culture. Your ability to do this will depend on a wide variety of factors. How much does the board already know about cyber threats and protection? How willing are they to make changes to improve outcomes?

The colloquialism "read the room" comes into play when joining a group of powerful people with their own ideas and agendas. When you align with their principles, it makes it much easier to ensure that your efforts are in harmony with their overall strategic objectives, risk management practices, regulatory compliance requirements, and comfort level.

Regulatory Compliance – The last things any corporation wants to risk are serious fines or negative legal penalties because they failed to uphold regulatory rules. There are many associated with cyber security specifically. Many are industry-specific like HIPAA, and others focus on doing business in different parts of the world such as the GDPR.

Keep up to date with the latest regulations and laws that affect your organization. If you approach security conversations with the Board of Directors without taking these into account, you will put the company at serious risk and lose the respect of other members.

Corporate Soft Skills – No one is going to get invited to a Board of Directors without already establishing that they have sufficient soft skills to provide benefits for the organization. These include things like the ability to communicate clearly and actively listen to other members, a desire to forge professional relationships with mutual benefit, strategic thinking abilities, holistic, long-term outlooks, and the ability to influence and persuade others when it is necessary to promote effective security improvements.

Diverse Collaboration -- One of the most important skills for any board member is the ability to work seamlessly together with everyone else involved in the strategizing and decision-making processes. Everyone on the board will have their own ideas and interests that benefit what they are responsible for specifically. Adaptability and flexibility are essential for making sure that they all accept new and improved cyber security protocols and strategies.

Additional Training for In-house Business Procedures

If a cyber security specialist comes to a new organization as part of their Board of Directors, they cannot expect to make an impact if they cling to their old job's ideas or protocols. Take the time to incorporate additional training focused on the organization that you are now responsible for. Understand the unique practices and company culture to gain insight into how to introduce new security measures and integrate them into existing operations and workflows.

When you delve into the intricacies of these procedures, it becomes much easier to not only identify potential vulnerabilities and areas of risk, but it also helps formulate solutions that will be most acceptable to other board members, executives, and employees alike. This is a huge part of your ability to communicate effectively within the Board of Directors and get things done. Read on to learn more specific details about working with diverse executives and shareholders as an inside cybersecurity specialist.

Communicating to the Board as an Inside Cyber Security Professional

No matter how skilled and experienced you are as a cyber security professional, you will not be able to get anything done or enact any affective changes if you cannot successfully communicate with the Board of Directors.

Even after an invitation to the Board and whatever onboarding procedures take place, the other members who have been there longer may not be open to your input automatically.

Your new role as a liaison between the technical intricacies of cyber security and the strategic decisions made by the Board puts you in a sometimes-difficult position. Mastering the art of communication is essential for ensuring that the suggestions you make align with their overarching goals and priorities. You undoubtedly would not have gotten this far without sufficient soft skills that made you seem like the best choice for inclusion. Now it is time to maximize the efficacy of those skills as an inside security professional who wants to make a real difference in the organization's future.

Understand Your Audience – Tailor Communication to the Board

As a cyber security professional, you understand the vital role that other Board members play in guiding and overseeing both short and long-term strategies. Effectively conveying intricate technical details and risks to people who do not know the cyber world well presents a challenge. The first step to making this work is understanding your audience.

The next step is finding a way to use diverse communication skills to boost engagement, educate as needed, and encourage a stronger culture and strategy for improved cyber security.

Board Member Roles and Industry Knowledge and Expertise

In an ideal world, every modern organization would be controlled by a Board of Directors who understand and embrace the truth about cyber security and its dynamic nature. Unfortunately, this is not what you will find in all cases. Before you begin to suggest new strategies, take the time to get to know the individual board member roles. How much do they know about tech security? How welcoming or resistant are they to change?

Consider asking or researching the answers to these questions about what other Board members know about cyber security specifically.

- What specific expertise or background do you have in the cyber security world?

- Have any of the other board members previously worked in any tech-related fields and what expertise do they bring to the table?

- What other industries or sectors have board members previously worked in? This can help you create understandable comparisons or examples for future communication.

- Are there any board members with specific regulatory or legal compliance expertise?

- How do they stay informed about cyber security trends, emerging threats, and regulatory changes?

- Are there any specific and serious gaps in the board's understanding of basic security details that require targeted education or training?

- Are there any board members who are specifically resistant to the idea of proactive security measures were those who take the "what we have now is good enough" perspective?

These are not the types of questions that you can directly ask the board members in most cases. It will require some homework into their pasts, current opinions about the security and tech world, and how they have worked with these things in the past before you joined. It is important to understand the foundation of the existing Board of Directors before you try to make positive changes that protect the organization now and in the future. These things will help you communicate more effectively and get things done.

Overall Priorities, Objectives, and Concerns

After determining who knows what and who will be on your side of the cyber security issue, the next step is to figure out what the board's priorities, goals, and specific concerns are. This will help you determine the active state of things and better formulate answers to their questions and acceptable plans for the future.

Ask a lot of questions and do a lot of research about what security protocols are already in place and how the other board members feel about them. Find out what focus areas drive board-level decision-making processes. When you gain this type of insight into what they most care about individually and collectively, you can better tailor your communication efforts to address these key issues and align your strategies with top organizational objectives. All of this makes it so much easier to get the support you need from other members to make changes.

Communication Styles and Preferences

People are more likely to accept what you have to say if you say it in a way that they can understand more clearly and find generally more engaging and interesting. A large part of this has to do with how the Board of Directors organizes meetings and collective communication in general. Does it meet twice a year in a corporate center for an in-person discussion? Do all the members participate in weekly zoom meetings instead? Is there a lot of communication between individuals that is not necessarily shared with the overall board?

By understanding and adapting to the diverse communication styles within the board, you enhance engagement, foster collaboration, and more effectively convey the significance of cyber security details. When you join, you should immediately know about arranged meetings and schedule expectations. It may take some time to learn the specifics of individual members' preferences.

Use all your soft skills to work with their diverse personalities to facilitate an overall culture of acceptance and enthusiasm when it comes to improving organizational security. This will help you align your efforts to protect systems, networks, and data as well as, ultimately, the people involved with the company, its reputation, and its finances.

Explaining Security Risks to Non-Tech Leaders

Other board members may possess extensive expertise in various business domains but lack specialized knowledge in cyber security. As the specialist in the group, you are responsible for bridging the gap between tech details and business priorities in a clear and accessible manner.

How you do this will depend on the level of knowledge and openness to the topic. The following strategies help you convey risks, current statuses, and future goals in a way that empowers them to grasp the significance and prioritize security-related initiatives and investment.

Translating Technical Jargon into Layman's Terms

Avoid using jargon that only people involved in the tech industry will understand. Likewise, do not automatically use acronyms or abbreviations. While part of your responsibility includes educating non-tech leaders about these things, you cannot expect them to understand or remember all of them from the start. Jargon will alienate your audience and make them less likely to act on important cyber security plans.

Instead, use plain language and provide relevant examples and metaphors that will make tech topics more accessible to everyone.

Since you probably use jargon and acronyms easily, this may take some time to rework your presentations or clearer communication.

Presenting Metrics and Key Performance Indicators (KPIs)

A large part of your position on the Board involves relating risks to business impact in ways that everyone can understand. Align the metrics and KPIs you share with topics of interest to the other members so that they can more easily gauge the effectiveness of the investments they agreed to. Some of these may include incident response times, breach detection rates, compliance adherence, and more. It will help to use visual aids like charts and graphs to make the complex metrics clearer.

Use a Storytelling Approach with Real World Examples

Narrative often works better than a direct quoting of statistics or facts when it comes to helping non-tech leaders understand complex issues. Metaphors, symbolic stories, and genuine case studies can go a long way to illustrating the potential impact of security vulnerabilities. Use news stories of other organizations who experienced serious data breaches as warnings.

These types of narratives provide context and help the other Board members understand the relevance of cyber security in every part of company operations and strategic planning. It will also help you get their attention rather than have them tuning out to jargon-filled speeches filled with data points and vague threats.

Provide Actionable Strategies and Recommendations

Once you have established sufficient understanding of cyber security weaknesses, threats, and potential outcomes in general, the time has come to offer solutions. Transition from discussing theoretical aspects to practical and actionable recommendations.

Be as clear as possible and tailor all suggestions to the company culture and operational style.

Focus on Practical Solutions for Risks in the PANCCD Pyramid

Remember the resilient cybersecurity hierarchy shown in the PANCCD pyramid when suggesting solutions that can eliminate vulnerabilities and improve response plans. Focus on pragmatic, hands-on, and budget-friendly solutions related to people, apps, networks, computing devices, code, and data protection. Make sure that the Board has all the tools, information, and guidance necessary to implement solutions that will work.

Create a Resilience Roadmap and Step-by-step Action Plan

No cyber security strategy is simple. These multifaceted roadmaps involve multiple steps that all lead to improvements. As the security specialist in the Board of Directors, it is your responsibility to identify key areas of vulnerability, assess potential risks, and develop a structured plan of action to reduce them and improve outcomes across the board. Additionally, the framework should include other things talked about in this book such as building a rapid response team, establishing protocols, setting up employee training, and strategizing about the potential need for public relations follow-up if something does go wrong.

Prioritize High-Impact Initiatives that Make an Immediate Difference

Security improvements involve long-term strategies and ongoing efforts if an organization wants to keep things as safe as possible forever. However, prioritizing changes or initiatives that have immediate impact is an essential part of communicating security needs to the Board. When you identify critical vulnerabilities, you must push for fast improvements. These may include things like turning on improved firewall protection, enforcing complex password rules, using multifactor authentication, and setting up automated threat detection systems.

Building Trust and Credibility Over Time

With the ever-changing and highly active world of cyber-criminal behavior, you must have what it takes to inspire non-tech leaders to enact immediate change when necessary. However, as a new member of the team, it may be difficult to get everyone on board. Real change may require you to build trust and credibility over time in order to get more support, allocate more resources, or make more sweeping changes to the existing security set up.

Cyber security professionals must demonstrate their commitment to protecting the organization's assets and improve the reputation by delivering results that matter. The more confidence you inspire, the more likely you are to get the support needed to improve security across the board.

Consistent Delivery to Demonstrate Reliability

Nothing speaks louder than results. This is an essential part of the process not only if a breach occurs or some other serious consequences happen. Maintain consistency in your message, intention, guidance, and improvements. The sooner you establish a reputation for reliability, the stronger their trust will grow. It is especially important to take a proactive stance toward security and then share the outcome of these efforts in alignment with top business interests.

Build Relationships and Cultivate Trust

Do not only show up for Board meetings and give quality demonstrations. Work to foster relationships with other members, talk with them about both cyber security topics and their own focus, and network effectively. This also helps you understand which members are more open to change and forward-thinking, tech-related innovations.

Maintain Transparency and Proactive Updates

Tell them exactly what you do, what improvements are made through those actions, and what you are doing next to boost the overall security standing of the organization.

Never hide security incidents in an attempt to make your efforts look more effective.

Transparency and accountability will go a long way to establishing trust. It will also encourage Board members to increase investment and include threat and vulnerability mitigation techniques into overall business strategies.

Demonstrate Results and Highlight Concrete Impacts

Nothing builds trust and credibility better than demonstrating exactly how your cyber security suggestions and initiatives improved organizational operation, prevented regulatory issues, or saved money in the long run.

How you present all the necessary information to enact security changes depends on the structure of the Board of Directors meetings and other specifics. For example, an in-person corporate meeting may give you the opportunity to share visual aids or digital presentations. Virtual meetings are much more common these days, with things like Zoom calls and videoconferencing standard practice.

No matter what options or limitations the specific communication style has, you must align your efforts to the Board's needs and interests.

When articulating the risks, pointing out the important KPIs, and addressing concerns, remember that one of the primary interests of any leadership group is profit. Every possible risk from interrupted business operations to customer data breaches all funnels down to financial loss. In the next section, you will learn more specific information about how to work with the Board of Directors in a way that brings their worst fears to light and supports their greatest desire.

Quantifying Cyber Risks from a Financial Perspective

As mentioned earlier, most Boards of Directors and companies themselves have a primary focus on financial matters. Every threat and vulnerability that affects day-to-day operations, reputation, or customer data carries a huge monetary risk. As cyber threats become increasingly sophisticated and prevalent, the potential consequences increase even more.

Understanding the Cost of Cyber Incidents

As a professional on the Board of Directors, it makes sense to approach every possible and actual incident in monetary terms. This helps ensure that other members understand what is at stake so that they allocate resources effectively and make smart decisions to mitigate future impact.

Direct Costs Explained

The direct costs of a cyber security incident depend on the type of organization, its operational style, how it makes money, and what type of breach or access issue occurs. This list outlines many of the common outcomes that do affect finances overall.

- Loss of revenue due to business disruption while the threat is active, or the recovery process is underway
- Immediate actions necessary to counter the incident may require additional people or work hours
- Costs of physical security to stop immediate access to computing devices and networks
- The need to create a business continuity plan or update cyber security strategies
- Increased insurance charges if the policyholder needs to pay out

- Greater expenses associated with talented recruitment since they may not want to work for a company with a history of cyber incidents
- Loss of intellectual property that affected overall organization income and standing
- Investment in risk assessment before the company resumes normal activity
- Additional costs with reestablishing or re-creating websites or other online platforms
- Legal costs associated with privacy violation claims or any other corporate attorney consolation services to recover from the attack
- License fees for new and improved security systems, software, and more

Indirect Costs of Cyber Security Problems

The indirect costs of a data breach or other security issue may occur very quickly after the problem arises or will continue to affect the company's ability to earn revenue far into the future. This depends largely on the type and severity of the incident as well as how much it affects the general public.

For example, a successful DDOS attack that brings down the company website for a day will have a much smaller impact than a network breach that steals healthcare information from thousands of consumers.

The indirect costs of a data breach include the potential:

- Decline in future revenue due to operational disruption
- Increased insurance costs or an inability to get the type of insurance that your organization needs
- Loss of productivity or smooth operations if you have to let personnel go or restructure your hiring plan

- Long-term need for investment in additional cyber security technologies, systems, training, and more
- Reputational damage that affects your entire brand's standing in the industry or niche
- Loss of foreign and domestic investors who do not want to be associated with that type of risk
- Increased regulatory oversight that can interrupt regular operations and cause ongoing expenses and concerns
- Losses in the stock market for publicly traded corporations

These are the types of things that cyber specialists must convey to the Board of Directors to get sufficient action and investment pointed at the entire security plan. When all of these direct and indirect risks exist, it is important to clearly discuss budget and resource allocation to prevent them.

Budgeting and Resource Allocation for Cyber Risk Management

Effectively communicating the need for more money to the Board of Directors may present various challenges depending on the individuals' interests and own projects. You must become an advocate for sufficient budget allocations and resource commitments both immediately and for future improvements. Always stress how threats and weaknesses can crop up and change at any time. Therefore, it is important to budget for ongoing efforts. When determining how to budget for cyber security and risk management, the PANCCD pyramid offers an exceptional model to follow.

People – Invest in Training

As the apex of the cyber security pyramid, people deserve sufficient investment to minimize human error, reduce risk of social engineering attack success, and bolster resilience across the board. Comprehensive training programs offer both immediate and long-term improvements. This type of professional development can also increase job satisfaction and help employees achieve professional goals, which makes it easier and more affordable to keep top talent.

Apps – Budget for Quality Software

An organization that has ignored cyber security for too long may have to upgrade apps and software systems that they use every day to meet security goals. Even those that focus on the cutting edge of tech advancements should still allocate resources to boost efficacy. Consider comprehensive assessments, more robust firewalls, automated updates and patches, and more.

Networks – Paying the Price for Security

Assessing network security both digitally and physically is the first step. Then, the call for increased budget or additional resources must focus on problem areas that introduce weaknesses. This can include everything from a biometric scanner on the server room door to the installation of intrusion detection systems and advanced encryption technologies.

Computing Devices – Quality Hardware and Firmware

Out of date hardware can wreak havoc on an otherwise secure system. Discuss the possibility of upgrading desktops, laptops, tablets, and phones issued by the company with the Board of Directors. If this is not necessary or feasible with the allowed budget, focus on firmware upgrades and the installation of things like endpoint detection and response systems, antivirus software, and device encryption tools instead. Device monitoring minimizes response times if something does happen, which can save a lot of money down the road.

Code – Oversight and Top Talent

Smart development practices lead to increased cyber security. These come from utilizing top talent in the programming field, which usually comes at a premium cost. Other places to encourage resource spending include static and dynamic application security testing, external code review platforms, and premium frameworks that focus on security. Talent plus the right tools gives the best and most resilient outcome.

Data – Risk, Protection, and Recovery

The foundation of the PANCCD pyramid represents the asset most at risk from cyber-attacks. Convincing Board members to spend more to protect customer and client information, intellectual property, proprietary business data, and more should be easy. Some smart investments include data encryption, more advanced access control, data loss prevention solutions, and a focus on regulatory compliance. Automated data backup and recovery solutions are must-haves in the case of a breach.

Understanding the financial implications of cyber security incidents makes it possible for the Board of Directors to make smart decisions about immediate and long-term efforts. Your ability to quantify cyber risks will provide valuable insight into the potential impact of any existing or emerging threat.

Through the analysis of various financial metrics including both direct and indirect costs, you enable them to prioritize investments effectively and protect their ultimate goal of increased profits.

The practice of integrating financial considerations into the overall risk management framework also helps board members speak more effectively with key stakeholders and customers. This leads to an overall increase in trust and reputation. All these things work together to encourage the Board of Directors to implement better cyber security measures, take a proactive approach to mitigating risks, and empower them to deal with attacks on valuable assets more quickly.

Mapping the FAIR Model to PANCCD

While individual organizations create their own strategies for understanding and dealing with cyber security risks, there is only one international standard model accepted today. The FAIR model stands for Factor Analysis of Information Risk. It helps organizations understand, analyze, and quantify risk from a financial perspective.

Learn about the FAIR model to help you as a guidepost on cybersecurity risk quantification.

> **Tip: Due to its common use and acceptance, FAIR (see https://www.fairinstitute.org) is a powerful tool for discussing cyber security resource allocation, budgeting, and strategy with the people who make decisions for your organization. It helps get everyone on the same page, clearly approaches decision-making, and demonstrates how money and time impacts the overall security status. It integrates well with other popular frameworks and is thus helpful for all organizations who want to improve their security profiles.**

How Does FAIR Actually Work?

In the simplest terms, this methodology looks at specific data points associated with cyber security issues, inputs them into the proprietary algorithms, and calculates risk based on how much money an organization could lose. It essentially multiplies the frequency of a loss event (how many times it could happen within a predetermined period) by its magnitude (the impact and scope of the cyber incident's outcome).

- Identify what risks are most likely to happen due to current and emerging threats in the cyber security realm

- Figure out how frequently these risks may result in loss for your organization. This is a matter of likelihood comparing threat frequency with details about vulnerabilities

- Determine how severe the losses would be for this type of cyber event. Take into consideration both direct or primary costs and secondary or indirect ones

- Use the FAIR model to calculate the overall risk value in monetary terms

Not only does this model for calculating cyber risk impact give organizations the type of information they need to take effective action, but it can also encourage smarter strategizing by looking directly at finances. Most Board members and shareholders are primarily concerned with the profitability of the company they represent or rely on.

Although complex, the information gleaned from FAIR provides a clearer picture of a variety of risk scenarios and helps leaders make smart decisions about how to mitigate and respond to them.

FAIR in Conjunction with PANCCD

This book's cyber security pyramid model gives organizations a framework to use something like FAIR in order to quantify every risk factor and incident possibility. Just as the different levels of the pyramid define all needed areas of focus when creating an effective strategy for business protection, they can also segment modeling efforts and provide structure to help those in charge better understand risks, potential losses, and the types of efforts they must make to improve.

People – The Human Factor

The FAIR cyber security model can help assess risks associated with human behavior, social engineering attacks, malicious actions, and errors to a high degree of accuracy. This includes the employee behaviors and their susceptibility to cutting corners or taking risks. When you can quantify the potential financial impact of these things, the organization can create a more effective strategy and system to minimize their effect.

When modeling these issues, you must look beyond employee ranks to outside human factors like contractors and third-party vendors. By using this systematic approach to assessing risks, it provides a more robust overall picture of every facet of people-centric security. The end result – prioritizing investments in things like training and behavioral analytics tools – goes a long way toward minimizing risk.

Apps – Software and Security

The assessment provided by the FAIR model also works with this layer of the PANCCD pyramid. Use it on both internally developed applications and third-party software solutions for maximum efficacy. This will primarily address vulnerabilities or flaws present in the applications and how they will impact the overall financial stability and success of the organization.

Using this method, you can identify and quantify potential consequences of exploitation, system downtime, and app-related reputational damage. Knowing the full scope of associated costs can help the company prioritize its efforts toward application security.

Networks – Infrastructure Risks

The infrastructures responsible for transferring data and managing digital communication within an organization need as much attention as any other part of the cyber security pyramid. Risk assessments with financial focus using the FAIR model will focus on breaches, data loss, service disruptions, and even the possibility of physical manipulation of the network hardware.

The ultimate outcomes of modeling risk include improvements to access controls, network monitoring capabilities, and advanced security tech. Knowing short and long-term financial risks fuels the company's desire to take a more proactive stance toward threat mitigation.

Computing Devices – Hardware and Firmware

The physical components that allow employees and other permitted users to access networks, apps, and data need as much security as any other level of the PANCCD pyramid. It is also possible to use the FAIR model of assessment to judge potential monetary effects of malicious actors taking advantage of weaknesses. When you know the likelihood and potential impact of these threats, it becomes much easier to fight against them. Endpoint security, firmware updates, protection against unauthorized access, and more all deserve smart investment and resource allocation.

Code – Development Options

The next layer of the cyber security pyramid focuses on code and, more specifically, the software development practices used within the organization or required during outsourced projects. Figuring out the financial impact of insecure code and resultant software vulnerabilities is possible using the FAIR model effectively.

The type of knowledge gleaned from these assessments makes it possible for companies to change anything within the software development lifecycle that offers nefarious actors a chance to affect or take advantage of the code. When you can identify, assess, and mitigate risks from the start, the potential impact decreases considerably. This is one of the most important aspects of overall cyber resiliency.

Data – The Foundation of PANCCD

The foundation of the security pyramid and the target of most hackers and malicious actors needs special attention when it comes to determining the current status of cyber protection and the potential risks associated with its failure. Using the FAIR model, you can gain valuable insights into how things like data breaches, losses, and manipulation can impact both direct and indirect costs now and in the future.

By analyzing a variety of factors surrounding the likelihood of an issue and the overall value of the data, this model makes it possible to accurately portray the threat landscape. Quantifying outcomes from a financial perspective does more to encourage change and investment than any theoretical information. By integrating FAIR into your risk management processes, you can identify and address the most serious and pressing security vulnerabilities. The end result is a much safer, resilient, confidential, and protected world of data.

In any conversation about cyber security, whether it is with the Board of Directors of a corporation, directly with shareholders, through marketing avenues with consumers, or with employees, clearly identifying risks and threats and the potential outcome matters. While robust assessment models like FAIR offer incredible options, the ability to improve security and work true resiliency into the organization depends on everyone's understanding that proactive and holistic strategies work best to protect your most valuable assets.

Complying with State, Federal and Global Regulations

In the realm of cyber security, successful implementation of effective strategies depends not only on technological advancements and operational efforts. It also must align with the complex web of regulations and compliance standards on the state, federal, and global level. No matter what the plans you put into place to protect your organization and what response protocols you develop, they must all comply with legal requirements if you do not want to run afoul of serious repercussions.

Understanding and adhering to these regulations is an essential part of establishing a strong cyber security plan for the future.

As your company fortifies its digital defenses against hackers, malicious software, and other nefarious actions, the legal mandates about safeguarding sensitive data, protecting consumer privacy, and ensuring digital integrity require your attention. First, understand the regulatory landscape as a whole and why it is so important to follow all the rules closely. Then, familiarize yourself with the specific cyber security regulations that apply to your operations, location, and industry specifically.

Overview of the Regulatory Landscape

No matter what industry or niche your company operates within, you know the importance of following all necessary laws and regulations that apply to it. These range from legal matters to physical safety rules. There are also many related to cyber security and digital protection of data and the overall infrastructure used to transmit it and communicate with the public.

Create a Compliance Team

One beneficial option that helps with compliance is the creation of an in-house team responsible for staying up to date with regulatory changes and ensuring that your organization follows them perfectly. The world of cyber security law is complex, and it takes a variety of individuals with expertise in tech, law, IT governance, and risk management to navigate it successfully.

The responsibilities of the regulatory compliance team will include:

Current Knowledge of the Rules – The team needs to know what rules and regulations to follow to create an effective strategy. It helps to have legal representation as part of this group.

Monitoring Potential Developments – The ever-changing landscape of security oversight means that there are frequent proposals or updates in the work from regulatory organizations. As with all other cyber threat strategies, it is important to take a proactive stance to regulatory compliance. That way, your organization will not be left playing catch up and potentially running afoul of the law.

Interpreting Regulations – Not only should the compliance team know the letter of the law, but also be able to apply it effectively to your organization's operations. Compliance sometimes looks different depending on the industry or type of data that your company works with. This is a matter of accuracy and effectiveness rather than trying to cut corners.

Compliance Plans and Activities – The team must create effective plans to comply with all necessary regulations and properly report as the government or industry organizations require. This process often includes an ongoing system of audits and assessments to track the overall posture of security efforts.

Assessing Operational Impact – How do the activities related specifically to regulatory compliance affect day-to-day operations and ongoing growth strategies? The compliance team should communicate with other in-house groups and also look outward toward industry trends in order to come up with the most seamless and non-interruptive practices.

Responding to Issues – With all the above planning involved in the cyber security compliance team's responsibility list, the risk of noncompliance is slim. However, you should always have a plan in place if something goes wrong. Create and enforce proper response procedures to minimize the legal, reputational, and financial impact of errors or oversights.

Stay Informed About Changes

Cyber threats constantly evolve, and new challenges emerge all the time in the global digital landscape. Not all governments or security organizations keep up to date on a minute-by-minute standard. You are primarily responsible for doing so to keep your protection strategy effective. By following the PANCCD pyramid levels and all the recommended efforts and protocols outlined in this book, you can do a good job of promoting continuous protection.

When it comes to staying informed about changes in the regulatory landscape, the existence of a compliance team makes sense. However, small or medium-sized enterprises or those not integrated fully into the technological world, may not have the resources to do so. Ignorance of the law is not an excuse for breaking it. Likewise, the groups that create and uphold security regulations will not offer leniency if you do not comply.

It is ultimately your responsibility to stay informed with every single new rule, regulation, law, and best practice on the state, federal, and global level. This is especially important for organizations that operate within certain industries, such as medical and healthcare, that control highly sensitive personal information or those related to finance or trade secrets.

Negative Outcome of Noncompliance

Whether your organization cuts corners on purpose or simply makes a mistake when it comes to complying with state, federal, or global regulations, the outcome can be severe and long-lasting. They may include any combination of the following:

Legal Penalties – One of the first and expected outcomes involves legal or regulatory penalties. These can include fines, sanctions, and legal action against the corporation or people involved in its leadership. These range in severity and have a widely disparate effect on the company's bottom line.

Reputational Damage – Any news of noncompliance may destroy the trust that consumers or other businesses have in your company. Partners and stakeholders will likewise struggle to regain their trust. This all becomes worse when it is publicized, and you do not have an appropriate marketing or public relations response prepared. Everything from customer loyalty loss to investors pulling their funding offers may happen.

Business Disruption – In some cases, lack of compliance with cyber security regulations may force a shut down or limit operations until the problem is fixed. These issues more often occur if you are faced with regulatory investigations and audits. Remediation efforts can divert resources and attention away from core business activities.

Loss of Market Opportunity – Noncompliance with certain regulations may limit the organization's ability to participate in diverse markets or industries that require close attention to the rules. Besides disrupting your overall operational strategies, this can also get in the way of opportunities for growth into new markets, expansion of product and service offerings, and an advance of competitive advantage.

Absolute Destruction – If the noncompliance issue is severe or long-lasting enough, it may result in a forced shutdown of the entire company. Any of the above-mentioned negative outcomes can lead to destruction. Excess legal costs and court judgments for those affected by the breach can add up quickly. Loss of reputation if the customers' personal, financial, or medical information is breached will destroy any trust in your organization. Downtime or operational disruption can greatly affect revenue. These are worst-case scenarios.

State Regulations: Compliance Requirements and Challenges

The complex landscape of cyber security laws and regulations begin at the state level for organizations that operate within the United States. These impose specific requirements and present unique challenges that must be addressed to mitigate legal and financial risks.

State regulations usually back up federal laws and align quite closely with international requirements as well. They are extra layers of protection for consumers and the companies that serve them. Always check with your state government and regulatory groups to determine the latest rules and enforcement mechanisms.

They may require additional paperwork or scheduling of audits and assessments to work into your overall cyber threat mitigation strategy.

Two of the most common and often-followed state regulations include:

California Consumer Privacy Act

The CCPA began in 2018 as a way to offer additional protection to California consumers. It covers their right to know if a company collected their information, the ability to opt out of this collection, and the right to delete personal data at any time. An organization must disclose these things and make it easy for consumers to control their own information access.

There are specific limitations of what organizations the California Consumer Privacy Act applies to. Your compliance team should investigate things like gross annual income, consumer data-driven revenue, and consumer data use to understand the requirements better.

Colorado Privacy Act

In 2023, the CPA became another popular state regulation for consumer data protection. Like the California rules, this primarily focuses on disclosure of data collection and usage as well as general security for the people themselves. Also, there are limits to the types of companies that must comply based on revenue levels or number of consumers whose information was collected.

Ultimately, it is up to a specific organization to stay informed about any new state regulations that apply to their place of business. Recognize that these do not only cover companies with headquarters in a specific state but also those who do business across state lines. Many of the rules and regulations are quite similar and simply good operational practices. Compliance always makes sense to avoid the potential negative outcome.

Federal Regulations: Understanding Standards

Interestingly enough, there are very few cross-industry federal regulations for cybersecurity in the United States. Other than the Homeland Security Act and those specifically related to sectors like healthcare, finance, and government operations, there are no general protections for consumer data or similar things. The specifics for industry-related regulations are covered below.

Some of the rules fall short of specifications for Internet-related organizations specifically or have vague language about cyber security protections. Phrases like a "reasonable level" of security are used, which can make it difficult for organizations to determine effective compliance measures.

It is nevertheless important to keep an eye on the federal US government's activities to know exactly what a company must implement and when. There are several proposals such as the Consumer Data Security and Notification Act to amend existing rules in place.

Global Regulations: Navigating International Compliance

Other countries around the globe offer more specific cyber security rules and regulations that should be followed by companies in the United States as well. These are required especially if your organization does business across national borders. In today's global marketplace, this is quite common, so it makes sense to stay informed and work toward full compliance.

The General Data Protection Regulation (GDPR)

This European regulation governs both the EU and any organization that collects, uses, or processes European citizen personal data. It began in 2018 and changed the landscape of international business operations, Internet use, and more. This has very strict data protection and personal information privacy requirements across industries. Noncompliance or any breach of the rules can lead to considerable fines and other repercussions.

Other International Cyber Security Regulations

The European Union Agency for Cybersecurity (ENISA) is the agency in that part of the world that makes regulatory recommendations, creates and enforces policies, and upholds compliance around the world. There are also protection rules and cyber-crime punishments in 156 other countries including major hubs of business such as Japan, Indonesia, India, Australia, Hong Kong, and Singapore. Some of these may require rule compliance, but many of them simply outline repercussions in case of data breaches or cyber-criminal activities.

For corporations that operate globally, the need to stay up to date with the latest laws and rules is paramount to success. Things can change very quickly, and the last thing any organization needs is to run afoul of international regulations. If you intend to expand into a new geographic market, take the time to investigate these things fully before finalizing the growth strategy.

Industry Specific Security Regulations Are More Common

Most of the cyber protection regulations an organization must follow relate to specific industries or types of data collected and used in-house or for standard business operations. Even if your specific industry is not represented in the collection of acts or rules listed here, that does not mean no regulations exist. This is a matter for investigation by your compliance team to ensure absolute legal operations and protections.

Health Insurance Portability and Accountability Act (HIPAA)

Organizations who operate in the healthcare and medical industries need to comply with all HIPAA regulations. Established in 1996, it has grown and changed slightly over the years to include privacy rules for patient personal information, necessary security measures to ensure absolute confidentiality, and strict notification protocols for notification of any breaches.

The related Health Information Technology for Economic and Clinical Health Act (HITECH) focuses on protection for electronic health records as new tech emerges and is adopted on a widespread basis. It promotes improved cyber security efforts, increases violation penalties, and requires prompt notification of any data breaches.

The Graham-Leach Bliley Act for Financial Institutions

This is one of the most widespread legislations designed to protect consumer financial privacy. It applies to any finance organizations or institutions including banks, credit unions, investment firms, and related things. The GLBA guidelines require annual privacy notices given to customers, specific security protocols that protect against unauthorized access, and third-party oversight to ensure multifaceted security.

Payment Card Industry Data Security Standard (PCI DSS)

Any company that accepts credit card or debit card transactions must comply with this standard. It protects cardholder information in both personal and financial rounds. Some of the regulations point to required encryption levels, network security, and auditing schedules.

Fair and Accurate Credit Transactions Act (FACTA)

This is another act that protects consumer credit information and card data. It is specifically focused on the prevention of identity theft and covers things like how much of a card number can be displayed on receipts or how to delete information that an organization no longer needs.

Government Sector Acts

These security regulations include the Federal Information Security Management Act (FISMA), the Cyber Security Information Sharing Act (CISA), and the Homeland Security Act of 2002. These are specifically for organizations that work with in or adjacent to the federal government.

Children's Online Privacy Protection Act (COPPA)

Introduced in 1998 and amended in 2013, this set of rules protects children under the age of 13 from privacy violations online. It involves such things as parental consent, privacy notifications, and a variety of other data protection practices.

Telecommunications Act of 1996

While much older and mostly focused on regulation of the telecommunications industry at large, it also includes certain rules specifically focused on ensuring safety. From a cyber security perspective, the main goals involve network access, interconnection, and integrity.

This is not a complete and exhaustive list of all acts, laws, rules, regulations, or best practices in the world of cyber security. As mentioned earlier, it is up to the individual organization or controlling members to understand and comply with everything that affects operations.

This is not a comprehensive list of every single rule and regulation related to cyber security across all industries. The head and members of an organization's security team and Board of Directors need to work together to ensure full compliance with everything related to the industry, niche, operational style, location, and more.

Compliance Strategies and Best Practices for Cyber Security

With all the different regulations that require attention, an organization needs to take special care when aligning their efforts and operations with requirements. The following steps or elements will help you devise methodologies and best practices that will streamline your compliance efforts while fortifying cyber security defenses.

It is important to note that regulatory compliance affects and intersects with every level of the PANCCD pyramid. From people and apps to computing devices and code, each part of the organization's function, task management, and future strategies must be taken into account.

Develop Policies and Procedures for Compliance

Executives, shareholders, team leaders, IT security specialists, and the compliance team must work together to create an effective collection of policies and procedures associated with legal regulations. Outline the rules, guidelines, and protocols that control the handling, protection, management, and use of sensitive and personal data. These things must be set in stone and cover every possible use case and issue that may arise when handling this type of information. Ultimately, regulatory compliance should automatically exist if the organization creates a strong cyber security strategy from the start.

Training and Awareness Efforts

Make sure everyone responsible for compliance efforts understands their role, what they need to do to meet regulations, and the risks associated with ignoring them or making extreme errors. Training and education are an essential part of getting people on board with all security plans. Stressing the importance of complying with these rules is even more important than suggesting safe methods for everyday tasks.

Continuously Monitor, Assess, and Report

As with all aspects of cyber security, regulatory compliance requires an ongoing effort to ensure you are following all the necessary rules and have the best protections in place. Set up digital monitoring of controls in order to identify gaps or weaknesses that may lead to issues with the controlling organizations. Also, regularly audit and assess the effectiveness of your strategies and protocols. Finally, stress the importance of 100% accurate and up-to-date reporting. The governments or groups that demand compliance will not give any leeway when it comes to reviewing any breaches or data handling discrepancies. Your organization should remain as strict.

Manage Vendor and Third-Party Risks

It is more difficult to control compliance associated with third-party vendors, platforms, or other external services like cloud-based networks or communication methods. It remains the company's responsibility to ensure that whoever you choose to work with takes legal matters as seriously as you do internally. Keep track of their policies and reports about any risky behavior or breaches.

Incident Response and Audit Preparation

Many cyber security laws have to do with responding to breaches or informing the public or affected consumers if someone had access to their information. Build a comprehensive incident response plan into your overall regulation strategy. This will not only help you effectively mitigate and manage security issues when they occur, but it will also keep you on the right side of the law. This can prevent additional fines or legal difficulty that can destroy your business reputation or operational progress.

Leveraging D&O, E&O, Keyman and Cyber Insurance

A large part of protecting an organization from cyber security issues plans on the shoulders of the various insurance policies held. The main types of these specifically designed to protect executives and the company itself offer not only financial coverage in case of a serious breach or other attack, but they also help provide peace of mind that your security strategy is complete.

Businesses rely heavily on technology and the expertise of specific individuals to maximize security against a wide range of cyber threats. As important as it is to have careful plans and protocols in place for everything from detection to response, it is impossible to prevent 100% of issues especially in a large corporation that handles valuable data and assets.

Proactive efforts are necessary, but it is simply a bad idea to ignore reactive options that help with recovery. These types of insurance policies are necessary in today's increasingly dangerous world.

Directors and Officers (D & O) Insurance

D & O insurance is a type of liability coverage that provides financial protection for directors and officers if they are sued for wrongful acts. Hopefully, the heads of a company are above reproach and do not cut corners or ignore certain rules, regulations, and good business practices. Even with a stellar reputation and close attention to detail, if a cyber security incident occurs, lawsuits or other personal expenses may happen.

Not only does this protect existing directors and officers of a specific organization, but it also helps attract the best to the company. If you are a cyber security specialist invited to join a Board of Directors or sign on to a specific corporation's top team, ask about D & O insurance coverage for your own personal protection. This is not a sign that you are prone to mistakes or oversights. It is something that all quality employers should have in place in this time of volatile cyber security challenges.

Some small or medium-size businesses may wonder if directors' and officers' insurance is right for them. Ultimately, it comes down to the structure of the company and whether these roles exist. A tech startup usually does not have a Board of Directors to worry about. This insurance should be kept in mind as the company grows.

Personal Protection for Directors and Officers

This type of insurance policy protects their personal assets in the case of judgment or other compensation. These executive leaders are hired to manage the company because of their skills, expertise, experience, and enthusiasm. They serve slightly different functions, which can change based on the specific organization they work for.

Directors are often chairman of the board or internal and external board members. Officers make up the C-suite of the executive level. They include the Chief Executive Officer (CEO), chief financial officer (CFO), and other roles like chief information officer, chief technology officer, and more.

D & O insurance protects personal assets. Policies may also integrate with business ones to offer more comprehensive coverage when responsibility is shared between the individual and the organization. If the director or officer is accused of fiduciary breaches, mismanagement, negligence, or legal or regulatory breaches, they will not be forced to pay with their own money. The insurance policy covers legal fees, judgments, and settlements. However, it is important to note that these insurance policies do not cover things like illegal actions, fraud, personal profiteering, any claims that exist prior to the policy creation, or liability issues like property or personal damages.

Errors and Omissions (E & O) Insurance

As the above insurance policies protect executives, E & O insurance also offers professional liability for professionals that are either employed by the organization or adjacent to it through outsourcing contracts and similar situations. They include lawyers, consultants, accountants, engineers, architects, or even cyber security specialists.

Since there is less direct control of their actions in relation to company operations, Errors and Omissions insurance policies are vital to the overall protection.

These policies are sometimes called professional liability insurance depending on the policy provider. When investigating options, it helps to look for coverage specific to your industry and the related cyber threats. If the part of regular operations includes consulting, advisory services, or similar things, E & O insurance is a must have.

What Do E & O Policies Cover?

This type of insurance is very similar to D & O insurance in that it covers legal costs, settlements, and judgments arising from legal claims or lawsuits. These focus on, as the name implies, errors, omissions, negligence, and failure to perform services as promised. These all integrate with potential breaches of contract as well as other specific instances.

Having this type of policy offers extra protection for professionals and businesses. The coverage handles financial losses related to alleged mistakes or oversights. No matter how tightly you run your operational ship, unwanted situations occur, and mistakes happen. Getting the right insurance policies can make all the difference for immediate recovery and long-term success.

Keyman (or Keyperson) Insurance

This is another type of insurance coverage that specifically focuses on executives or high-value employees that offer undeniable contributions to the organization's success. Instead of focusing on lawsuits or other legal actions, keyman insurance is a type of life insurance policy. If the individual passes away or becomes disabled and thus unable to fulfill their business duties, this insurance policy will kick in to help support the business during the transitional period.

Keyman Insurance Policy Coverage

When a company employs a high-value individual who offers immeasurable help with management, operational strategies, cyber security protection, or other duties, recovering after their death presents many challenges. This is quite different from instances where an executive or employee decides to move on to another job position or retires. Both of those instances usually come with a considerable amount of warning that allows the company to prepare.

If a catastrophic event occurs, the organization will struggle to not only replace the individual but take up the slack from their responsibilities and duties. Both operational and financial loss can occur during this time. The insurance policy can be used to cover costs associated with finding a replacement, training and onboarding, compensation for lost revenue, repaying debts, or supporting business operations in a variety of other ways.

It makes sense to leverage the power of a keyman insurance policy to protect the business from serious and potentially devastating impact of losing an individual whose skills, knowledge, experience, and network are crucial to its success and stability.

Cyber Insurance

From a cyber security perspective, it makes sense to look into cyber insurance, which is otherwise known as cyber liability insurance. These policies are designed to protect businesses from financial loss and liability resulting from data breaches and other incidents. These can also include ransomware attacks, D DOS attacks that lead to business disruption, direct hacking, and other cyber threats.

Vulnerabilities and threats exist at every level of the PANCCD cyber security pyramid. No matter how comprehensive the protection plan and security strategies are, bad things can still happen. Having insurance coverage just in case can help alleviate the horrible outcome and get the organization back on track as quickly as possible.

Expenses Covered by Cyber Insurance

In short, any cyber-attack can lead to additional expenses not only for the company's recovery efforts but also due to liability toward business partners, consumers, and regulatory groups. This insurance covers the following depending on the specific policy details:

- Data Breach Response Costs – Investigating, notifying, monitoring, and recovering from any data breaches takes money, often in the form of additional compensation for the people who are responsible for doing these important tasks. And

- Cyber Ransom and Extortion – If the only solution to a ransomware or similar attack is to spend the money to regain access, the insurance policy will cover it.

- Data Recovery and Restoration – It costs money to work through the systems and recover data and restore it to the most recent uncorrupted form. This will help the organization get back to business as quickly as possible.

- Legal and Regulatory Expenses – If the data breach violated laws or regulatory compliance rules, or if a lawsuit occurs on behalf of contacts or consumers, a cyber insurance policy will step in and cover regulatory investigations, fines, penalties, and judgments.

- Business Interruption – Most serious cyber security events interrupt normal operations for a period of time. These lead to loss of income and extra expenses to get things moving again. The insurance helps prevent longer or more serious interruptions that can lead to reputational damage and loss of revenue.

Leveraging Insurance Strategically for Cyber Security

Purchasing insurance policies is not enough to fully realize the power of their coverage. It takes strategic use to mitigate risks and enhance the overall resilience of a company. Of course, the protection they afford depends largely on the specific policies purchased and maintained. Take special care to find the best for the specific organization, executives, high-value employees, and cyber security position.

How do you do this? As with every other aspect of security strategizing, choosing the right insurance policies begins with conducting a thorough risk assessment. Dive into specifics to find potential areas of vulnerability and exposure. Investigate every director, officer, executive, and consultant to determine both threat factors and potential disruption should they suddenly become unavailable. When you understand the specific risks, it becomes much easier to tailor insurance coverage to address them comprehensively.

Leveraging insurance policies also involves integrating them into the broader risk management framework the organization uses. Develop clear policies and procedures for incident response, which should include notifying policy holders and all necessary paperwork or communication with coverage providers. Ensure that the rapid response team and other employees are trained and equipped to handle potential threats. Make sure that they and especially the individuals covered by the insurance understand it fully. This helps align policy coverage with evolving risks.

Despite the importance of D & O, E & O, keyman, and cyber insurance, it is not necessary to pay the highest price to get needed coverage. Take the time to actively engage with insurers and brokers to negotiate favorable terms and conditions specific to the organization's operations. In most cases, bespoke policies that protect specific individuals or situations will work better and cost less than comprehensive plans.

Finally, the presence of diverse insurance coverage demonstrates the overall commitment to cyber risk management and best practices when it comes to business governance, executive oversight, employee protection, and risk management as a whole. While the average consumer may know nothing about the insurance policies you carry, the people associated with day-to-day operations and future strategies will. They help instill confidence in your ability to manage risks effectively and mitigate potential financial losses if cyber-attacks and data breaches occur.

Companies and other organizations who are serious about their cyber security protection strategies need to invest in diverse and comprehensive insurance policies. Leveraging these options gives you not only increased financial security but also peace of mind that the company can recover more quickly and efficiently from any horrible situation.

The overall complexity of cyber security, efforts and strategies necessary to protect organizations, and the diverse responsibilities of everyone from the top directors to the lowliest employees cannot be overstated. No matter what role you play in a company's success, today's rapidly evolving world of cyber threats will affect your efforts.

It takes a diverse, dynamic, and robust plan to minimize risks, reduce the chance of operational disruption, maintain full compliance with all necessary rules and regulations, and protect the business and individuals from serious repercussions.

With full understanding of the PANCCD pyramid, and the vulnerabilities and responsibilities at every level – people, acts, networks, computing devices, code, and data – organizations are better equipped to approach the realm of tech protection with more assurance that their efforts will be successful. This is the way to create truly resilient cybersecurity.

PANCCD™ Incident Response Plan

The **PANCCD™ Model** (People, Apps, Networking, Computing, Code, and Data) provides a structured, actionable approach to managing and mitigating cybersecurity incidents. By addressing each core element, this framework ensures a holistic response that minimizes damage, restores operations, and fortifies defenses against future attacks. Here's a five-step incident response plan aligned with PANCCD principles.

1. Preparation
Preparation is the foundation of any effective incident response plan. This stage involves building a robust security posture that leverages the PANCCD pillars:

- **People**: Train your team to recognize and respond to potential threats, ensuring everyone understands their roles in incident response.

- **Apps**: Regularly assess and patch applications to close vulnerabilities that attackers might exploit.

- **Networking**: Implement network segmentation and monitoring tools to detect anomalies early.

- **Computing**: Ensure endpoints are equipped with advanced security tools, such as EDR (Endpoint Detection and Response).

- **Code**: Conduct routine code reviews and vulnerability scans to minimize software flaws.

- **Data**: Encrypt sensitive data and maintain secure, frequent backups to facilitate recovery.

Comprehensive preparation allows for faster, more coordinated responses when an incident occurs.

2. Detection and Analysis
The detection phase involves identifying potential incidents through monitoring tools, user reports, or automated alerts. PANCCD principles guide the analysis:

- **Networking**: Use intrusion detection systems (IDS) and traffic logs to pinpoint unusual activity.

- **Data**: Evaluate logs and event data to identify the scope and impact of the incident.

- **Apps and Code**: Analyze affected applications and scripts for signs of compromise.

Rapid detection ensures timely responses, while thorough analysis provides insights into the incident's root cause, affected systems, and potential vulnerabilities.

3. Containment, Eradication, and Recovery

Once an incident is confirmed, immediate steps are taken to limit damage, remove threats, and restore systems:

- **Containment**: Isolate affected systems, such as disconnecting compromised devices or restricting access to breached accounts.

- **Eradication**: Remove malware, unauthorized access points, or malicious actors from the environment. This step involves analyzing code and applications for backdoors or hidden threats.

- **Recovery**: Restore systems using clean backups, ensuring no malicious elements are reintroduced. Verify that all patches and updates are applied.

By addressing all PANCCD components, this phase ensures a swift return to normal operations with minimized residual risk.

4. Communication

Clear, timely communication is crucial during and after an incident. This includes internal and external stakeholders:

- **People**: Notify executives, IT teams, and affected employees about the situation and provide instructions.

- **Networking**: Collaborate with ISPs or third-party providers if network infrastructure is impacted.

- **Apps and Data**: Inform customers, partners, and regulators if sensitive data or critical applications are affected.

Transparent communication builds trust and ensures compliance with legal and regulatory requirements.

5. Post-Incident Review and Improvement

After resolving an incident, conduct a comprehensive review to identify lessons learned and prevent recurrence:

- **People**: Evaluate team performance and update training programs to address gaps.

- **Apps and Code**: Identify and patch vulnerabilities discovered during the incident.

- **Networking and Data**: Update monitoring tools and strengthen encryption protocols.

The post-incident review transforms the event into an opportunity for growth, enhancing the organization's overall cyber resilience. The PANCCD™ Incident Response Plan provides a robust framework for addressing cybersecurity incidents. By integrating all six pillars—People, Apps, Networking, Computing, Code, and Data—this model ensures a well-rounded, proactive approach to threat management. Whether handling a minor breach or a major incident, PANCCD equips organizations with the tools to respond effectively, minimize damage, and build stronger defenses for the future.

What to Do Immediately After a Breach

A cybersecurity breach is one of the most challenging situations an organization can face. How you respond in the first hours and days is critical to mitigating damage, restoring operations, and rebuilding trust. The PANCCD™ Model provides a structured framework for addressing the chaos of a breach while ensuring a calm, calculated response.

Key Steps to Respond Effectively to a Breach

1. Stay Calm and Activate Your Incident Response Plan

- **Purpose**: Avoid knee-jerk reactions and follow a structured approach.

- **Action Steps**:
  - Gather the incident response team, led by the CISO or equivalent.

o Review the predefined incident response plan and assign roles immediately.

2. Contain the Breach

- **Purpose**: Prevent further damage by isolating the threat.

- **Action Steps**:
  o Disconnect affected systems from the network to stop the spread.
  o Disable compromised accounts or endpoints immediately.

3. Preserve Evidence for Forensics

- **Purpose**: Ensure a proper investigation by keeping records intact.

- **Action Steps**:
  o Avoid altering or shutting down affected systems unless necessary.
  o Secure logs, snapshots, and system data for forensic analysis.

4. Engage Key Personnel and Secure Their Roles

- **Purpose**: Ensure the CISO and their team are supported to lead recovery.

- **Action Steps**:
  o Offer six months of job security to the CISO and key cybersecurity staff to focus on addressing the breach and rebuilding defenses.
  o Hold regular check-ins to track progress and ensure team morale.

5. Contact Authorities and Third Parties

- **Purpose**: Comply with laws and involve appropriate support.

- **Action Steps**:
    - Notify law enforcement (e.g., FBI Cyber Division) if the breach involves criminal activity or national security threats.
    - Inform regulatory authorities as required (e.g., GDPR, CCPA, HIPAA).
    - Engage third-party incident response firms if internal resources are insufficient.

6. Communicate Transparently

- **Purpose**: Maintain trust with stakeholders through honest and timely updates.

- **Action Steps**:
    - Notify senior leadership, legal counsel, and PR teams immediately.
    - If customer data is affected, issue a statement outlining the breach and the steps being taken to mitigate risks.

7. Eradicate the Threat

- **Purpose**: Remove malicious actors and secure systems.

- **Action Steps**:
    - Conduct a root cause analysis to identify vulnerabilities.
    - Patch systems, remove malware, and block unauthorized access points.

8. Restore and Recover Operations

- **Purpose**: Safely resume business processes.

- **Action Steps**:
  - Restore systems using clean, tested backups.
  - Monitor restored systems closely to ensure no hidden threats remain.

9. Conduct a Post-Breach Analysis

- **Purpose**: Learn from the incident to prevent future breaches.

- **Action Steps**:
  - Review logs, forensics, and response efforts to identify gaps.
  - Update security policies, procedures, and training programs based on lessons learned.

10. Implement Long-Term Improvements

- **Purpose**: Build a stronger cybersecurity foundation for the future.

- **Action Steps**:

  - Strengthen the PANCCD pillars:

    - **People**: Increase security awareness training.

    - **Apps**: Audit third-party applications for vulnerabilities.

    - **Networking**: Strengthen segmentation and access controls.

- **Computing**: Patch systems and enhance endpoint protection.

- **Code**: Conduct regular application security testing.

- **Data**: Implement encryption and secure backup policies.

Why the PANCCD™ Model Matters in a Breach

The PANCCD™ Model provides a holistic, step-by-step approach to breach response, addressing every critical aspect of cybersecurity. By focusing on containment, collaboration, and continuous improvement, it helps organizations recover quickly while minimizing long-term risks. A breach isn't the end—it's an opportunity to rebuild stronger, learn, and stay ahead of future threats.

The PANCCD™ Model for Digital Executive and Family Protection

In today's interconnected world, digital threats target not just organizations but also individuals and families. Executives face heightened risks, including identity theft, cyberstalking, and privacy breaches that can extend to their loved ones.

These threats are further magnified by vulnerabilities in IoT devices, smart technologies, and digital services used daily.

The **PANCCD™ Model**—focusing on People, Apps, Networking, Computing, Code, and Data—provides a comprehensive framework for safeguarding yourself, your family, and your digital presence at home and while traveling.

People: Awareness and Preparedness

Human behavior is often the weakest link in security. Awareness and proactive preparation are key to minimizing risks.

- **Educate Your Family**: Teach family members to recognize phishing attempts, avoid oversharing on social media, and report suspicious activity. For children, emphasize the dangers of sharing personal information online.

- **Develop Emergency Protocols**: Create a plan for responding to lost devices, hacked accounts, or suspicious activity. Ensure every family member knows what to do in these scenarios.

- **Training for Executives**: Enroll in executive protection courses that address specific risks such as targeted cyberattacks or corporate espionage.

Apps: Securing Digital Tools

Applications used on personal devices and smart home systems can become gateways for hackers. Securing them is vital.

- **Enable Multi-Factor Authentication (MFA)**: Protect accounts with MFA to ensure an extra layer of security beyond passwords.

- **Audit Permissions**: Regularly review and limit app permissions to reduce access to sensitive data.

- **Avoid Unnecessary Apps**: Remove unused or untrustworthy applications from all devices.

- **Parental Controls**: Use trusted parental control apps to monitor children's activities and block inappropriate content.

Networking: Fortifying Connections

A secure network is the backbone of digital safety at home and on the go.

- **Segment Your Network**: Separate IoT devices, like smart cameras and thermostats, from your primary devices by creating a guest network.

- **Secure Wi-Fi**: Use strong, unique passwords and enable WPA3 encryption. Change default SSIDs to something non-identifiable.

- **Virtual Private Network (VPN)**: Use a VPN for all internet activity, particularly on public Wi-Fi, to encrypt communications and mask browsing activity.

- **Monitor IoT Traffic**: Use apps like Fing to scan your network for unauthorized devices or unusual activity.

Computing: Securing Devices

Hackers target the devices we use every day, from laptops to gaming consoles. Ensuring their security is non-negotiable. Beyond antivirus and firewall, you will need to:

- **Enable Encryption**: Protect your devices with full-disk encryption to secure stored data if lost or stolen.

- **Use Secure Smart Devices**: Choose baby monitors, wireless doorbells, and other IoT devices with built-in encryption and advanced security features. Disable remote access when not needed.

- **Keep Firmware Updated**: Regularly update the firmware on all IoT devices to patch vulnerabilities.

- **Remote Wipe Capabilities**: Configure devices to allow remote wiping in case of theft.

Code: Protecting Software Integrity

Software vulnerabilities are a primary attack vector for hackers targeting both personal and professional environments.

- **Secure Custom Apps**: If using custom software for personal or executive needs, ensure regular security audits.

- **Disable Unnecessary Features**: Turn off risky features like Universal Plug and Play (UPnP) on IoT devices, which can expose them to external threats.

- **Browser Protection**: Use secure browsers with anti-tracking and ad-blocking extensions to prevent malware infections and data collection.

Data: Safeguarding Personal Information

Your data, both online and offline, is a prime target for identity theft and privacy breaches.

- **Limit Data Sharing**: Avoid providing unnecessary personal information to apps, websites, and services.

- **Encrypt Sensitive Data**: Use encrypted cloud storage for sensitive documents like IDs, medical records, and financial statements.
- **Regular Backups**: Back up critical data securely to prevent loss from ransomware attacks or hardware failures.

- **Monitor Identity**: Consider subscribing to identity protection services like LifeLock to receive alerts about potential misuse of your personal data.

Hidden Threats in Everyday Life

Hackers are increasingly exploiting vulnerabilities in everyday devices and situations. Here's how to address specific risks:

1. **Baby Monitors and Doorbells**

   o **Risk**: Hackers can hijack live feeds to spy or communicate.

   o **Solution**: Use models with encryption and disable remote access when not in use.

2. **Hidden Cameras**

   o **Risk**: Rental properties or hotel rooms may contain concealed recording devices.

   o **Solution**: Perform visual and RF detection scans to identify hidden devices. Use apps to find unusual Wi-Fi-connected devices.

3. **Smart Devices Like Alexa**

   o **Risk**: Eavesdropping on private conversations.

   o **Solution**: Mute microphones when not in use, limit permissions, and delete stored recordings regularly.

4. **Gaming Platforms**

   o **Risk**: Phishing, account theft, and exposure to inappropriate content.

   o **Solution**: Enable parental controls, use strong passwords, and teach children secure gaming habits.

What to Do If Your Identity Is Stolen

If your identity is compromised, immediate action is essential to mitigate damage:

1. **Contact Credit Bureaus**: Place a fraud alert or freeze your credit with Equifax, Experian, and TransUnion.

2. **Notify Financial Institutions**: Inform your bank and credit card companies to block unauthorized transactions.

3. **File a Report**: Submit an identity theft report to the FTC and your local police department.

4. **Dispute Fraudulent Activity**: Work with creditors and credit bureaus to remove unauthorized accounts.

5. **Monitor Continuously**: Use identity monitoring services to track further suspicious activity.

Security While Traveling

Travel introduces unique risks, including physical theft and unsecured networks.

- **Pre-Travel Prep**: Limit the devices and sensitive data you bring. Use temporary devices when traveling to high-risk regions.

- **Hotel and Rental Safety**: Check rooms for hidden cameras and secure personal belongings. Avoid discussing sensitive topics over hotel Wi-Fi.

- **Social Media Discipline**: Post about your travels only after returning home to reduce risks of targeting.

The PANCCD™ Model offers a comprehensive framework for digital executive and family protection. Security is not just about technology; it's about peace of mind for you and your family.

> **Tip: By proactively securing devices, networks, and personal information, you can safeguard yourself and your loved ones from evolving threats.**

Conclusion: Why the PANCCD™ Model is Your Key to Cyber Resilience and Regulatory Compliance

The PANCCD™ Model—focused on People, Apps, Networking, Computing, Code, and Data—offers a comprehensive and actionable framework to build cyber resilience while achieving regulatory compliance.

By simplifying the complexities of cybersecurity, it empowers organizations to act quickly and effectively against evolving threats.

Here's why the PANCCD™ Model is your essential tool:

1. **Comprehensive Coverage**: It ensures every critical element of your digital ecosystem is protected, addressing vulnerabilities across all domains—People, Apps, Networking, Computing, Code, and Data.

2. **Simplified Risk Management**: By breaking cybersecurity into manageable pillars, the model makes it easier to identify risks, prioritize solutions, and deploy safeguards.

3. **Supports Regulatory Compliance**: Its holistic approach aligns with key regulatory frameworks like GDPR, CCPA, and HIPAA, streamlining compliance efforts and reducing legal risks.

4. **Quick Wins for Resilience**: The model emphasizes impactful, immediate actions that strengthen your defenses without requiring extensive resources or overhauls.

5. **Adaptability to Emerging Threats**: PANCCD™ equips you to respond swiftly to new challenges, ensuring your organization remains resilient in a dynamic threat landscape.

6. **Ease of Implementation**: Designed for both novices and experts, the framework provides a clear roadmap that is straightforward to understand and execute.

7. **Enhanced Stakeholder Confidence**: By safeguarding data and systems effectively, the PANCCD™ Model builds trust with clients, partners, and regulators alike.

8. **Future-Proof Security**: It creates a solid foundation for long-term protection, helping your organization stay one step ahead of the next breach.

💡 **Tip: Leverage the PANCCD model using commonsense. Keep it simple, focus on improving weakest areas of highest risk, first, and then keep improving, while documenting progress at improving resiliency for proof of due care and due diligence and to allow the organization to take necessary business risks and to be more profitable.**

By adopting the PANCCD™ Model, you'll not only protect your organization from current risks but also ensure compliance, strengthen trust, and future-proof your cybersecurity strategy—all in a quick, efficient, and practical way.

You will never be 100% secure but you will be more resilient and ready, reducing risks, complying with regulations, improving productivity and adding value to your organization's bottom line.

www.ingramcontent.com/pod-product-compliance
Lightning Source LLC
Chambersburg PA
CBHW070400200326
41518CB00011B/1999